THE MORAL
ECONOMY OF
MOBILE PHONES

PACIFIC ISLANDS
PERSPECTIVES

THE MORAL ECONOMY OF MOBILE PHONES

PACIFIC ISLANDS PERSPECTIVES

EDITED BY ROBERT J. FOSTER
AND HEATHER A. HORST

Australian
National
University

PRESS

PACIFIC SERIES

ANU PRESS

Published by ANU Press
The Australian National University
Acton ACT 2601, Australia
Email: anupress@anu.edu.au
This title is also available online at press.anu.edu.au

A catalogue record for this book is available from the National Library of Australia

ISBN(s): 9781760462086 (print)
9781760462093 (eBook)

Cover image: Woven string bag (wool-based yarn), artist unknown, 2015. Eastern Highlands Province, Papua New Guinea. Photo by H. Horst.

Cover design and layout by ANU Press

Contents

Discussion

List of Figures

Acknowledgements

The chapters in this volume, with the exception of Holly Wardlow's contribution, were originally presented at the 2015 meeting of the American Anthropological Association in Denver. We thank the authors for their effort and patience in making both the original conference session and this volume possible.

An Australian Research Council Discovery Project grant (DP140103773, The Moral and Cultural Economy of Mobile Phones in the Pacific) awarded to the editors provided material support for this volume as an open-access publication. We thank Stewart Firth for considering the volume as part of the ANU Press Pacific Series. Two anonymous reviewers offered constructive feedback on the chapters, and Justine Molony copyedited the chapters with professional care. We are grateful to Katie Taylor and Bernard Prasad for assistance with several images, and to Digicel Group Limited for kind permission to reproduce images in three of the chapters. While Digicel Group has permitted the use of its images and artwork, the views, opinions and research expressed in this volume are those of the authors and do not necessarily reflect the official policy or position of Digicel Group or any of its affiliates and entities.

Contributors

Robert J. Foster is Professor of Anthropology and Visual and Cultural Studies, and Richard L. Turner Professor of Humanities at the University of Rochester. His research interests include globalisation, corporations, commercial media and material culture. He is the author of *Social Reproduction and History in Melanesia: Mortuary Ritual, Gift Exchange, and Custom in the Tanga Islands* (Cambridge University Press, 1995); *Materializing the Nation: Commodities, Consumption, and Media in Papua New Guinea* (Indiana University Press, 2002); and *Coca-Globalization: Following Soft Drinks from New York to New Guinea* (Palgrave Macmillan, 2008). His most recent book is *Art, Artifact, Commodity: Perspectives on the P.G.T. Black Collection* (Buffalo Society of Natural Sciences, 2015; co-edited with Kathryn H. Leacock).

Heather A. Horst is Professor of Media and Communications at the University of Sydney. Her research focuses upon understanding how digital media, technology and other forms of material culture mediate relationships, communication, learning, mobility and our sense of being human. Her co-authored and co-edited books examining these themes include *The Cell Phone: An Anthropology of Communication* (Berg Publishers, 2006); *Living and Learning with New Media: Summary of Findings from the Digital Youth Project* (MIT Press, 2008); *Hanging Out, Messing Around and Geeking Out: Kids Living and Learning with New Media* (MIT Press, 2009); *Digital Anthropology* (Berg, Publishers 2012); and *Digital Ethnography: Principles and Practice* (Sage Publications, 2016).

Margaret Jolly (FASSA) is a Professor in the School of Culture, History and Language in the College of Asia and the Pacific, The Australian National University. She was an Australian Research Council (ARC) Laureate Fellow 2010–2015, and has written extensively on gender in the Pacific, exploratory voyages, missions and contemporary Christianity, maternity and sexuality, cinema and art; she is currently researching gender and climate change in the Pacific, funded by an ARC Discovery Project. Her most recent book is *Gender Violence and Human Rights: Seeking Justice in Fiji, Papua New Guinea and Vanuatu* (ANU Press, 2016; edited with Aletta Biersack and Martha Macintyre); her full list of publications is available at researchers.anu.edu.au/researchers/jolly-ma

Dan Jorgensen is Associate Professor of Anthropology at the University of Western Ontario. His research interests include mobile telephony and social media, Christianity and transnational evangelism, and the anthropology of mining. He is the author of numerous articles including 'Who and what is a landowner? Mythology and marking the ground in a Papua New Guinea mining project' in *Anthropological Forum*; 'Third Wave evangelism and the politics of the global in Papua New Guinea: Spiritual warfare and the recreation of place in Telefolmin' in *Oceania*; and 'Mining narratives and multiple geographies in Papua New Guinea: Ok Tedi, the emerald cave, and Lost Tribes' in *Journal de la Société des Océanistes*.

Daniela Kraemer is an applied anthropologist focused on understanding the lived experiences of marginalised populations with the aim of using the data and insights to push for social change. She is lead ethnographer at InWithForward, a social service design shop, and teaches anthropology at Wilfrid Laurier University. Her research interests include urban practices of marginalised youth in Port Vila, Vanuatu, and experiences of social services by street-involved adults, isolated seniors and newcomers across Canada.

David Lipset is Professor of Anthropology at the University of Minnesota. He has done fieldwork in Papua New Guinea intermittently since 1981. He is the author of *Gregory Bateson: The Legacy of a Scientist* (Beacon Press, 1980); *Mangrove Man: Dialogics of Culture in the Sepik Estuary* (Cambridge University Press, 1997); and *Yabar: The Alienations of Murik Men in a Papua New Guinea Modernity* (Palgrave Macmillan, 2017).

Jeffrey Mantz is Program Director in Cultural Anthropology and Human Subjects Research Officer at the National Science Foundation, where he has served since 2012. He holds a PhD in anthropology from the University of Chicago and has previously taught at George Mason University, Cornell University, California State University at Stanislaus and Vassar College. His research takes him to the Caribbean and Central Africa, where he explores issues related to inequality, resource extraction and commodity supply chains.

Holly Wardlow is Professor of Anthropology at University of Toronto. Her research interests include gender, sexuality, interpersonal violence and medical anthropology. She is the author of *Wayward Women: Sexuality and Agency in a New Guinea Society* (University of California Press, 2006) and a co-author of *The Secret: Love, Marriage, and HIV* (Vanderbilt University Press, 2009). She is currently completing a sole-authored monograph about HIV/AIDS in Papua New Guinea.

Introduction

Robert J. Foster and Heather A. Horst

Digicel and the Mobile Revolution

Over the past decade, Pacific Islanders have witnessed remarkable growth in the presence and adoption of mobile telecommunications. For example, average mobile coverage (2G) in Fiji, Samoa, Solomon Islands, Tonga and Vanuatu jumped from less than half of the population in 2005 to 93 per cent of the population in 2014 (Pacific Region Infrastructure Facility 2015). A 2015 Groupe Speciale Mobile Association (GSMA) report notes that unique subscriptions across the entire region increased from 2.3 million in 2009 to 4.1 million at the end of 2014 – a 12.6 per cent annual growth rate. This 'mobile revolution' is one result of the liberalisation of the telecommunications sector that opened national markets to competition in the 1990s.

While at first liberalisation affected the provision of landlines and postal services, reforms in information and communication technology (ICT) policy and regulation, which were spearheaded by the World Bank, led to partnerships between global telecommunications companies, such as British-owned Cable and Wireless, and national carriers in the region. Eventually, the markets for mobile phone services were opened up. Vodafone entered the market in Fiji in 1994 and was the sole mobile telecommunications provider until 2008. In Papua New Guinea, however, the only operator until 2007 was state-owned Telikom PNG. Its subsidiary, Bemobile, had in 2006 only 100,000 mobile phone subscribers (Cave 2012: 5), most of whom were based in the capital, Port Moresby.[1]

---

1   In 2013, the PNG Government acquired 85 per cent of the shares of Bemobile Limited. In 2014, Bemobile signed a non-equity partner market agreement with Vodafone Group Plc., the world's largest mobile operator. The new entity operates in Papua New Guinea and Solomon Islands as bmobile-vodafone.

Elsewhere, Telecom Vanuatu focused on delivering mobile services on the main island of Efate and in the capital of Port Vila until a new carrier, Digicel, entered the market in 2008.

For many in the Pacific, the story of the mobile revolution begins with the arrival of Digicel Group Limited (Digicel). Digicel started operations in Jamaica in 2001 and expanded throughout the Caribbean and Central America over the next five years (Horst and Miller 2006). Founded by Irish businessman Denis O'Brien, the nimble mobile network provider is privately owned – 94 per cent by O'Brien himself – and registered in Bermuda with headquarters in Kingston, Jamaica. Digicel's global mobile subscriber base has grown from 400,000 in 2002 to 13.6 million across more than 30 countries (Digicel n.d.).

Strategically, Digicel targeted 'high risk' countries with fairly small populations in the developing world. They introduced aggressive competition into markets where telecommunications services were provided, often ineffectively, by state-sponsored monopolies. Digicel's interest in the Pacific Islands officially began with the award of a licence to operate in Samoa in 2006 and, soon afterwards, in Papua New Guinea, Fiji, Tonga, Vanuatu and Nauru. Digicel also attempted to obtain a licence in the Solomon Islands in 2007, but was blocked following a local effort to protect the state-owned incumbent, Solomon Telekom. After subsequent attempts to acquire a licence in 2008 and 2009, Digicel lost its bid to Bemobile.[2] (The company is not present in New Caledonia, which has a well-developed telecommunications infrastructure maintained by one mobile phone and 3G broadband carrier, OPT Telecom.)

In Nauru, a tiny island republic with 10,000 citizens, Digicel now claims 100 per cent of the mobile market (Digicel Group Limited 2015). Papua New Guinea is by far the company's largest and most lucrative Pacific market where Digicel has at various points held up to 97 per cent of mobile market share (Digicel Group Limited 2015). Vanuatu has approximately 180,000 connections for a population of 261,000 (GSMA 2015) and Digicel claims 69 per cent of its mobile market (Digicel Group Limited 2015). Somewhat surprisingly, given the relative economic health of the country, Fiji – where Digicel holds approximately 33 per cent market share (Digicel Group Limited 2015; the actual rate might be as low as 20

---

2   For maps of Digicel's markets in the Caribbean, Central America and the Pacific, see the company's website: www.digicelgroup.com

per cent) – remains one of its least successful operations. Vodafone Fiji continues to dominate the market due to its historical relationship with Fijian state agents and agencies as well as the goodwill associated with the company's philanthropic foundation. The country's skilled workforce and relatively comfortable standard of living have, however, made Fiji an ideal choice for the location of Digicel Pacific's regional headquarters, which are situated in downtown Suva. Overall, the Pacific region still has one of the lowest rates of penetration for mobile subscriptions in the world, including in its largest global market – Papua New Guinea – where subscriber rates linger around 41 per cent (Digicel 2015; other estimates put the rate at as low as 31 per cent [GSMA 2015]). Nevertheless, the perception of untapped market potential and future growth in mobile phone adoption motivates Digicel's continued investments in the region.

Across the six Pacific countries in which Digicel operates, there have been significant costs associated with investing in mobile markets. These costs involve negotiating with national land trust boards and customary landowners to place towers in the most effective locations for signal transmission; installing towers in remote regions, access to which often requires lengthy boat rides or the use of helicopters followed by a long hike for technicians carrying the parts to be assembled; and establishing a retail network of vendors to sell and distribute handsets, subscriber identity module (SIM) cards and prepaid airtime.

During the first five years of operations, Digicel managed to build goodwill and enjoy popularity among consumers in the region. The introduction of competition reduced the price of calls and the availability and accessibility of handsets. The establishment of mobile networks in poorly served regions enabled Digicel to reach more marginalised segments of the population by building a consumer base of low-income and rural communities. The company's self-proclaimed 'bigger and better network' allowed for stronger connections between family and friends in remote villages and the growing urban centres across Melanesia. Digicel also engaged in well-publicised acts of corporate philanthropy, such as providing aid after cyclones and contributing to education, youth, community and health initiatives. They established the Digicel Foundation in Papua New Guinea in 2008 and invested in the 'cultural' dimensions of their new consumer markets by sponsoring rugby and other popular sports teams as well as music competitions.

This initial goodwill, however, is increasingly contested as Digicel attempts to maintain its competitive edge and to reach out to new consumers while still dealing with the constraints of physical geography that have shaped mobile networks throughout the region. In many remote areas, 2G networks and the use of satellites limit the network speed and access available to consumers, who largely use prepaid voice and short message services (SMS). In urban areas, an increase in smartphones and the growing appetite for social media have spurred demand for faster speeds (4G and long-term evolution [LTE]) and provision of data services and bundles, such as day passes and free access to Facebook.

Digicel has embraced digital media convergence, evolving into a provider of content as well as connectivity. The company has diversified into the consumer entertainment market, launching its own television network and acquiring cable and satellite television broadcasters and Internet service providers across the Pacific region. States have accordingly revisited their relationship with the mobile network operator in response to changing licence requirements and regulatory environments.

***

It is not easy for anthropologists to keep up with the mobile revolution in the Pacific Islands, especially in the highly diverse countries of Melanesia, which the chapters in this volume highlight. Take, for instance, Papua New Guinea. By 2010, all of the billboards in towns and point-of-purchase signage in rural areas that had once advertised Coca-Cola soft drinks in red and white were replaced with publicity for Digicel. Only the colours remained the same (Figure 1). While researchers were just beginning to consider the opportunities that widespread use of basic handsets for voice calls and texting might afford people across the nation, the uptake of inexpensive smartphones subsidised by mobile network operators had already begun.

By 2016, there were some 800,000 smartphone users and 400,000 Facebook users in Papua New Guinea. The number of Internet users as a result had increased from under 2 per cent of the population in 2010 to over 11 per cent. And the potential for phone-enabled access to social media was becoming clearer. On 7 June 2016, the *Guardian* (Davidson 2016) reported in its US online edition that police in Port Moresby, Papua New Guinea's capital, had opened fire on university students protesting government corruption. The report contained a brief video clip, apparently recorded on a smartphone and uploaded to a Twitter account, in which gunshots could be heard. New technologies were making it possible for international news agencies to spread accounts of events in locales where few foreign correspondents tread and, perforce, for mobile phone users, acting as citizens, to redefine their relations to the state. This latter possibility is one example of what we have in mind when speaking of the moral economy of mobile phones.

Figure 1. Digicel billboard in Goroka, 2014
Source: Photo by R. Foster

Mobile Phones from a Moral Economy Perspective

The idea of a 'moral economy' acquires significance by contrast with an ideological notion of 'market economy' (see Thompson 1967, 1971). This notion of market economy refers to a world characterised as modern in which people meet each other as transactors and only as transactors – strangers who seek to maximise their own interests within the limits of custom and the law. Such transactions presuppose no prior relationship between the parties involved and entail no obligation to sustain a relationship once the transaction is complete. Friendship and kinship, loyalty and propinquity are all beside the point. Economic activity is disconnected from other considerations and other ways of reckoning exchange, and disembedded from social relations that might entangle and motivate buyers and sellers (Polanyi 1957).

Regardless of whether such a cold and impersonal economy exists, its force as a rhetorical image is undeniably real. On the one hand, anthropologists since Marcel Mauss have used the idea of such a market economy as a foil for the notion of a gift economy and for a more humane form of capitalism. On the other hand, the people that anthropologists sometimes study have themselves used the idea to imagine alternative and positively valued forms of tradition and custom. The aim of these critiques is simply to affirm that production and exchange, distribution and consumption all involve, inevitably and irreducibly, multi-stranded and durable relations among human beings inhabiting shared social circumstances.

For our purposes in this volume, the moral economy of mobile phones implies a field of shifting relations among *consumers*, *companies* and *state agents*, all of whom have their own ideas about what is good, proper and just. These ideas inform the ways in which, for example, consumers acquire and use mobile phones; companies market and sell voice, SMS and data subscriptions; and state agents regulate both the everyday use of mobile phones and the market activity of licensed competitors. Ambivalence, disagreement and ongoing negotiation about who owes what to whom are thus integral features of the moral economy of mobile phones.

Companies, consumers and state agents all impinge upon each other in ways that make it difficult to consider one apart from the others. Telecommunications companies may not operate without state-issued spectrum licences; consumers cannot operate their mobile phones without registering their SIM cards, a requirement that states oblige companies to enforce; and states count on the revenue of taxes and fees collected from companies. Nevertheless, as the chapters in this volume demonstrate, it is possible to bring one of the three actors into the foreground while temporarily placing the other two in the background.

Consumers

Looking at what consumers do with mobile phones has so far been the strength of the emerging anthropological literature based on ethnographic research in Melanesia (see, for example, Andersen 2013; Lipset 2013; Jorgensen 2014; Telban and Vávrová 2014; Kraemer 2015; Macdonald and Kirami 2015; Taylor 2015). The chapters in this volume extend this work, demonstrating how consumers navigate the possibilities and perils of creating novel forms of interpersonal intimacy at the same time that they purchase and exchange airtime (phone credit) with expectations and ideals derived from longstanding practices of reciprocity and kinship.

Mobile phone use entails both continuity and change in the practices by which people produce social relations. For example, the mobile phone has become an increasingly important tool for consolidating, expanding and managing social networks. Holly Wardlow notes in her chapter in this volume (Chapter 2) that women, in particular, have embraced mobile phones as a way to stay in touch with close natal kin after marriage into their husband's household. David Lipset (Chapter 1) convincingly demonstrates how rural residents of the Lower Sepik River area of Papua New Guinea put the mobile phone in the service of consolidating kinship across the physical distance separating town and village. A single handset, fortuitously located in one of the few spots where a connection to the Digicel network is relatively reliable, enables rural–urban communications that facilitate the discharge of moral obligations associated with reciprocal gift giving. Indeed, the owners of the handset make it available to other community members as a gift, declining to charge their neighbours a fee for using the phone and instead relying on occasional counter-gifts of airtime (credit) to keep the service active.

Similarly, Daniela Kraemer (Chapter 5) shows how young men in Port Vila effectively continue the practices of village 'big men' by accumulating relationships, adding contacts to their mobile phones in an effort to expand a personal network that can be drawn upon for support when necessary (see also Servy 2014). Yet, while there is undoubtedly continuity, there is simultaneously change, for young men attempt to create personal networks of support in a social context where they have diminishing access to the resources of kin in town and relatives in rural homes. Accordingly, these young men often seek to recruit strangers to their networks, calculating the marginal value of a new contact in terms of its potential as a source of future support. Such calculations can lead to the sort of rational cost–benefit analysis associated with market transactions: Is it worth it to me to respond to this request for credit, or to answer that call? Robert Foster (Chapter 6) demonstrates that this kind of calculation can involve relations with kin as well as strangers, and is a function more generally of managing the exigencies of prepaid phone subscriptions in which constant awareness of one's current balance is a practical necessity. Prepaid subscriptions can thus function as technologies for promoting forms of self-discipline and budgeting reminiscent of the personal ethics of Max Weber's ascetic Protestants.

Figure 2. bmobile-vodafone headquarters in Port Moresby, 2017
Source: Photo by R. Foster

It is perhaps the combination of intimacy and strangeness made possible by mobile phones that marks something new under the sun in Melanesia (and elsewhere; see, for example, Archambault 2013; Gilbert 2016). The random dialling in search of 'phone friends' and the uneasy, fantastic relationships that develop between persons who never meet face-to-face and who never know for sure the true identity of their interlocutors afford a new kind of pleasurable sociality (Andersen 2013; Jorgensen 2014). Affective intimacy is possible, but at a distance that protects the parties in the relationship from their vulnerability to each other's demands (often sexual demands placed on women by men). Wardlow's chapter examines an extraordinary example of such sociality in which Lucy, an HIV-positive woman marginalised by her own kin, received emotional comfort and material support from an intimate stranger, Angela, a woman living in a nearby town who responded to Lucy's random call with compassion. The case of Lucy and Angela dramatically reveals how mobile phones have been configured by some consumers as 'affective technologies': tools that enable people – especially women – to escape the constraints not only of physical distance, but also of gendered relations of power.

State Agents and Agencies

State agents and agencies shape the moral economy of mobile phones in numerous ways. The executive decision to open up national markets to foreign companies is often legitimated as a democratic effort to make telecommunications accessible and affordable to all people. Once this goal has been more or less achieved, state agents often attempt to regulate the various ends to which consumers put new technologies, such as the use of mobile phones for online consumption of pornography and dissemination of political critique through social media such as Facebook. In Papua New Guinea, SIM card registration became mandatory in early 2016. The legislation requires Digicel and bmobile-vodafone to collect and archive personal information about all SIM card owners at point of sale. The legislation was introduced along with a new cybercrime policy, the stated intention of which is to address the abuse of mobile phones in making harassing or fraudulent calls and the use of social media to transmit defamatory, 'false' and 'offensive' claims (Pacific Media Centre 2015).

In Fiji, SIM card registration became mandatory in 2010. According to a 2 July 2010 report from Radio New Zealand, 'The Attorney General, Aiyaz Sayed Khaiyum, says a significant number of telephone users are misusing the service, including making bogus and threatening calls, and bomb threats' (Radio New Zealand 2010). Nevertheless, critics of the regime that came to power in 2009 led by Commodore Voreqe 'Frank' Bainimarama saw the decree as part of a more extensive initiative to monitor telecommunications (see, for example, Hayes 2010). In Papua New Guinea, too, SIM card registration and the cybercrime policy promoted by the National Information and Communications Technology Authority (NICTA) attracted strong criticism. Former prime minister

Sir Mekere Morauta declared the new regulations a threat to freedom of speech and the freedom and independence of the media: 'These precious rights and freedoms are under attack as never before by a Prime Minister demonstrably determined to silence legitimate criticism, including criticism of the official corruption that appears to exist at the heart of government' (Pacific Media Centre 2015).

These concerns about unwanted and unwarranted state surveillance resonate with the anxieties expressed by Jorgensen's (Chapter 3) interlocutor, Toby, regarding the spectre of American and Chinese imperialism and the prospect of a one-world government. Toby interprets the cell towers installed by Digicel as signs and instruments of an impending apocalypse, the result of power exerted by foreign (corporate) entities beyond the control of the PNG state. Toby's close reading of the new 100 kina banknote, adorned with an image of a cell tower, reasonably poses the moral question of who indeed owns and controls the nation's telecommunications network. Liberalisation in Papua New Guinea – a response to the state's own failure to deliver telecommunications services to the mass of the people – effectively put large chunks of this network in the hands of a privately owned offshore corporation.

Toby's anxieties prompt consideration of the various ways in which state agencies collaborate with companies (some state-owned) in building and maintaining the material infrastructure that subtends mobile communications and information technologies. In Papua New Guinea, NICTA requires Digicel to extend its network into some rural areas where market considerations might not prove to be an incentive – a challenge that Digicel has more than met (even if some areas, such as the Murik Lakes described by Lipset (Chapter 1), remain without reliable connectivity). State agents – namely, individual members of parliament – can generate goodwill among their constituents by securing Digicel towers with the use of discretionary funds or by arranging for very small aperture terminal (VSAT) connections. Through such acts, MPs discharge their moral obligation to bring development in the form of goods and services to the people whom they represent.

Similarly, although Digicel has built an extensive network of cell towers throughout the country, it must still deal with the PNG state for access to the undersea fibre-optic networks that land at Port Moresby and Madang. The fees associated with such access, and the present state and future capacity of the undersea cables themselves, are currently issues of political contention. The plan to create a national transmission network (NTN) calls for a new state-owned entity, DataCo, that will function as the wholesale seller of bandwidth to retailers such as Digicel. But the plan hinges on Telikom, the state-owned company responsible for landlines, yielding control of the international cable gateways – one of its most valuable assets (see Oxford Business Group 2014).[3]

Figure 3. A Vodafone retail agent in Suva, 2017
Source: Photo by R. Foster

The NTN is publicly represented as a kind of moral action: the state's effort to meet the challenge of making mobile communication – in particular, access to broadband – affordable and available to all. In this regard, the NTN continues the more familiar work of the state regulator (NICTA in Papua New Guinea) responsible for maintaining a level playing field for competition in the telecommunications industry. Such interventions have included pressure on companies in Papua New Guinea to make their networks inter-operable and, in Fiji, to reduce the rates charged for voice calls made from one network to another. These interventions are sometimes undertaken by non-state actors who advance the cause of consumer citizenship by appealing to a notion of universal human rights. In Fiji, for example, the Consumer Council lobbies both companies and state agencies to implement policies and practices that protect and empower individuals for whom mobile connectivity has

---

3   In February 2017, the PNG Government announced the amalgamation of DataCo, bmobile-vodafone and Telikom under a new name, Kumul Telikom.

become a fundamental necessity (Consumer Council of Fiji 2014). In this sense, the mobile phone becomes – like clean water and access to healthcare – a sign and instrument of belonging in a modern world order. In such an order, unlike in Toby's dystopic universe, the rights of individuals command respect.

Companies

The flipside of consumer citizenship is corporate citizenship (see Foster 2008). Companies, like state agents, fashion themselves as positive moral agents through a variety of techniques. These techniques range from supporting highly visible public programs of corporate social responsibility to circulating even more visible advertising images on billboards, in newspapers and through online social media. By these means mobile network operators stake out their places in the moral economy vis-à-vis each other and the broader public of citizens and consumers, whose interests the companies address.

Heather Horst's chapter (Chapter 4) demonstrates how Digicel in Fiji competes with its well-resourced rival, Vodafone, in what might be regarded as a tournament of consumer citizenship that renders alternative claims to national service as rival marketing propositions. Vodafone Fiji, now a wholly locally owned entity, returns shareholder value to the Fiji National Provident Fund, which in turn pays dividends to the people. The company thus fulfils its obligation to share its profits with all Fijians (something that Digicel, as a privately owned company – mostly by one man who resides for tax purposes in Malta – cannot claim as plausibly). Citizenship moves into the foreground. Digicel, however, boasts lower rates and extensive coverage in rural areas long neglected by the state-owned telecommunications providers. Actual consumer benefits thus eclipse promises of national citizenship.

Advertising and marketing campaigns for mobile services reveal how corporate personnel imagine the communities that they seek to engage. In Melanesia, this imagination often unfolds in national(istic) terms (Foster 2005), such as in the duelling advertisements that Digicel and Vodafone launched in in Fiji in 2008 (Horst, Chapter 4). Missteps can occur, of course, as when some of Digicel's snappy youth-oriented advertisements were perceived as Caribbean imports unsuited to the conservative moral climate of Fiji. But sponsorship, philanthropy and disaster relief provide ongoing evidence of active participation in and commitment to the national communities in which companies operate. Digicel and Vodafone thus fashion themselves as super-citizens, willing and able to take on responsibilities that exceed their obligations as market actors.

Mobile marketing stimulates not only the collective imagination, but also, in the form of promotions, the consumption of individual mobile users. Many of these promotions, circulated through text messages sent sometimes several times a day (or night!), seek to engage consumers in an exchange relationship. In exchange for topping up airtime credit by a certain amount, for example, a consumer will receive 'free' talk-time to be used during a certain period of the day. Or consumers will be 'rewarded' with extra airtime credit if they top up on a particular day of the week. These promotions, which can be precisely targeted at consumers on the basis of the data that usage generates, suggest that the companies recognise their moral obligations to acknowledge the gift of patronage offered by their consumers. By the same token, these promotions, as well as other services that companies provide, such as the opportunity to receive a loan of airtime credit (at about 30 per cent interest), shape the moral economy of phone use. Individuals defer calls to certain times of day, or send text messages instead of making voice calls, as part of an effort to conserve and extend scarce resources (see Foster, Chapter 6). Promotions often exert a paradoxical effect on consumers, enticing them to spend in order to save; that is, to demonstrate their thriftiness through an increase in consumption. Some promotions, however, go further, enticing consumers to increase consumption through participation in contests for prizes that critics, such as Fiji's Consumer Council, liken to an unregulated form of gambling. In these cases, the consumer is imagined less as a partner in a relationship of mutual exchange and more as a desiring subject from whom revenue can be extracted.

Towards a Moral Economy of Mobile Phones in Melanesia

The anthropology of mobile phones has perhaps given more attention to how consumers use ('appropriate' or 'domesticate') a relatively new technology than to how companies and state actors enable and regulate such use (see, however, Doron and Jeffrey 2013). Unsurprisingly, ethnographic approaches to phone usage have revealed, if not celebrated, local variations in how the technology has been taken up, a function of variation in the social and cultural imperatives of the users themselves. Consider, for example, the unforeseen popularity of texting in the Philippines (Pertierra et al. 2002), 'link-up' in Jamaica (Horst and Miller 2005) and even 'phone friends' in Papua New Guinea (Andersen 2013) and Vanuatu (Kraemer 2015).

The chapters in this volume demonstrate similar instances of cultural creativity and rightly emphasise the agency of people such as Toby and Lucy and Angela in making mobile phones what they are in Melanesia. But the moral economy framework proposed here and developed in the following chapters also seeks to extend our view of the socio-material network that mobile phones bring into being. This network includes, at a minimum, the consumers who use the phones, the companies that supply consumers with handsets and airtime, and the state agents that regulate (even if haphazardly) both telecommunications companies and the infrastructure that subtends mobile phones. It is the effects of all the elements in this network – including the phones themselves as material objects with definite qualities – that at any given moment inform what an ethnographic approach can tell us about mobile phones in Melanesia or elsewhere. Put differently, while it is certainly important not to reduce phone usage to the automatic outcome of either corporate strategies or state policies, it is equally important to acknowledge that such usage happens in circumstances that users themselves have not chosen. The chapters in this volume respond to this double challenge.

In lieu of a conclusion, the book ends with the comments of Jeffrey Mantz and Margaret Jolly, who reflect upon the contribution of the chapters both to the anthropology of Melanesia and Oceania and to a general understanding of the processes that underpin the mobile revolution globally.

References

Andersen, B. 2013. Tricks, Lies and Mobile Phones: 'Phone Friend' Stories in Papua New Guinea. *Culture, Theory and Critique*, 54(3): 318–34. doi.org/10.1080/14735784.2013.811886

Archambault, J.S. 2013. Cruising through Uncertainty: Cell Phones and the Politics of Display and Disguise in Inhambane, Mozambique. *American Ethnologist*, 40(1): 88–101. doi.org/10.1111/amet.12007

Cave, D. 2012. *Digital Islands: How the Pacific's ICT Revolution is Transforming the Region*. Lowy Institute for International Policy. Sydney, Australia.

Consumer Council of Fiji 2014. *Issues Paper: Problems Faced by Consumers in the Mobile Phone Sector*. Suva, 31 March.

Davidson, H. 2016. Papua New Guinea: Four Students Reported Dead after Police Open Fire on March. *Guardian*, 7 June, www.theguardian.com/world/2016/jun/08/papua-new-guinea-police-shoot-at-students-during-march (accessed 1 December 2016).

Digicel Group Limited n.d. History. www.digicelgroup.com/en/about/history.html (accessed 15 April 2018).

—— 2015. Form F-1, Registration Statement. Securities and Exchange Commission. Washington, DC.

Doron, A. & R. Jeffrey 2013. *The Great Indian Phone Book: How the Cheap Cell Phone Changes Business, Politics and Daily Life*. Cambridge, Massachusetts: Harvard University Press.

Foster, R.J. 2005. *Materializing the Nation: Commodities, Consumption and Media in Papua New Guinea*. Bloomington: Indiana University Press.

—— 2008. *Coca-Globalization: Following Soft Drinks from New York to New Guinea*. New York: Palgrave Macmillan.

Gilbert, J. 2016. 'They're My Contacts, Not My Friends': Reconfiguring Affect and Aspirations Through Mobile Communication in Nigeria. *Ethnos*, 83(2): 237–54. doi.org/10.1080/00141844.2015.1120762

Groupe Speciale Mobile Association (GSMA) 2015. *The Mobile Economy: Pacific Islands 2015*. www.gsma.com/mobileeconomy/pacificislands/ (accessed 1 December 2016).

Hayes, M. 2010. Fiji – No Endgame in Sight. Webdiary, 10 August. webdiary.com.au/cms/?q=node/3097 (accessed 2 December 2016).

Horst, H.A. & D. Miller 2005. From Kinship to Link-up: Cell Phones and Social Networking in Jamaica. *Current Anthropology*, 46(5): 755–78. doi.org/10.1086/432650

—— 2006. *The Cell Phone: An Anthropology of Communication.* New York: Berg.

Jorgensen, D. 2014. *Gesfaia*: Mobile Phones, Phone Friends, and Anonymous Intimacy in Contemporary Papua New Guinea. Paper presented at CASCA: Canadian Anthropology Society Conference. York University, Toronto, 30 April.

Kraemer, D. 2015. 'Do You Have a Mobile?' Mobile Phone Practices and the Refashioning of Social Relationships in Port Vila Town. *The Australian Journal of Anthropology (TAJA)*, 28(1): 39–55. doi.org/ 10.1111/taja.12165

Lipset, D. 2013. *Mobail*: Moral Ambivalence and the Domestication of Mobile Telephones in Peri-Urban Papua New Guinea. *Culture, Theory and Critique*, 54(3): 335–54. doi.org/10.1080/14735784.2013.8265 01

Macdonald, F. & J. Kirami 2015. Women, Mobile Phones, and M16s: Contemporary New Guinea Highlands Warfare. *The Australian Journal of Anthropology (TAJA)*, 28(1): 104–19. doi.org/10.1111/taja.12175

Oxford Business Group 2014. Establishing Stable Links and Increasing Broadband Access are Priorities. www.oxfordbusinessgroup.com/ analysis/establishing-stable-links-and-increasing-broadband-access-are-priorities (accessed 8 July 2017).

Pacific Media Centre 2015. PNG: Former PM Criticises New Cybercrime Laws. 29 October 29. www.pmc.aut.ac.nz/pacific-media-watch/png-former-pm-criticises-new-cybercrime-laws-9467 (accessed 2 December 2016).

Pacific Region Infrastructure Facility 2015. Economic and Social Impact of ICT in the Pacific. Report summary. www.theprif.org/documents/ kiribati/information-communication-technology-ict/economic-and-social-impact-ict-pacific (accessed 14 April 2018).

Pertierra, R., U. Eduardo, A. Pingoi, J. Hernandez & N. Dacaney 2002. *TXT-ING Selves: Cellphones and Philippine Modernity*. Manila: De La Salle University Press.

Polanyi, K. 1957 [1944]. *The Great Transformation: The Political and Economic Origins of Our Time*. Boston: Beacon Press.

Radio New Zealand 2010. Fiji Phone Users Must Register Under New Decree. 2 July. www.radionz.co.nz/news/world/50623/fiji-phone-users-must-register-under-new-decree (accessed 1 December 2016).

Servy, A. 2014. 'Do You Have a Boyfriend? No, I Don't Have a Phone': Sexuality, Morality and Mobile Phones in Port-Vila, Vanuatu. Paper presented at CASCA: Canadian Anthropology Society Conference. York University, Toronto, 30 April. www.academia.edu/6954883/2014_Do_you_have_a_boyfriend_No_I_don_t_have_a_phone_Sexuality_morality_and_mobile_phones_in_Port-Vila_Vanuatu_CASCA_conference_Session._Melanesian_Promise_s_Uncertain_Prospects_Precarious_Relations_and_Hedged_Bets_in_the_Western_Pacific_30_04_2014_York_University_Toronto_Canada_ (accessed 15 April 2018).

Taylor, J.P. 2015. Drinking Money and Pulling Women: Mobile Phone Talk, Gender and Agency in Vanuatu. *Anthropological Forum*, 25: 1–16. doi.org/10.1080/00664677.2015.1071238

Telban, B. & D. Vávrová 2014. Ringing the Living and the Dead: Mobile Phones in a Sepik Society. *The Australian Journal of Anthropology (TAJA)*, 25(2): 223–38. doi.org/10.1111/taja.12090

Thompson, E.P. 1967. Time, Work-Discipline, and Industrial Capitalism. *Past and Present*, 38: 56–97.

—— 1971. The Moral Economy of the English Crowd in the Eighteenth Century. *Past and Present*, 50: 71–136.

1

A Handset Dangling in a Doorway: Mobile Phone Sharing in a Rural Sepik Village (Papua New Guinea)

David Lipset

Figure 4. Aimaru performed in Darapap village to welcome the author

Source: Photo by David Lipset, 2014

In August 2014, I hired a small boat to travel east along the 65-kilometre stretch of coast from Wewak town to the Murik Lakes, a rather remote and undeveloped region where I have done fieldwork since the 1980s. After three hours or so on the ocean, the boat entered one of the two openings into the lakes, which are a large coastal system of mangrove lagoons at the mouth of the Sepik River, and drew up to the lakeshore at Darapap village (see Figure 5).[1] A small welcoming ceremony was underway for which a handful of men and women had adorned themselves with shell ornaments and red body paint. Just below the front of the Men's House, they were performing Aimaru, the dance that features a fish-spirit figure on a pole encircling the troupe as its holder bobs it up and down (see Figure 4; see also Lipset 2009). On the boat, a villager from Wewak town, who was with me to witness the little spectacle, wondered out loud, in a tone of mild exasperation, whether someone had died. As we climbed out of the boat, a middle-aged woman, whom I had known for many years as an older sister of my ex-wife, came and took my hand and led me through the thick, wet mud to shake hands with everybody: the dancers, the chorus of men with hand drums, as well as a few spectators who had turned up to watch.

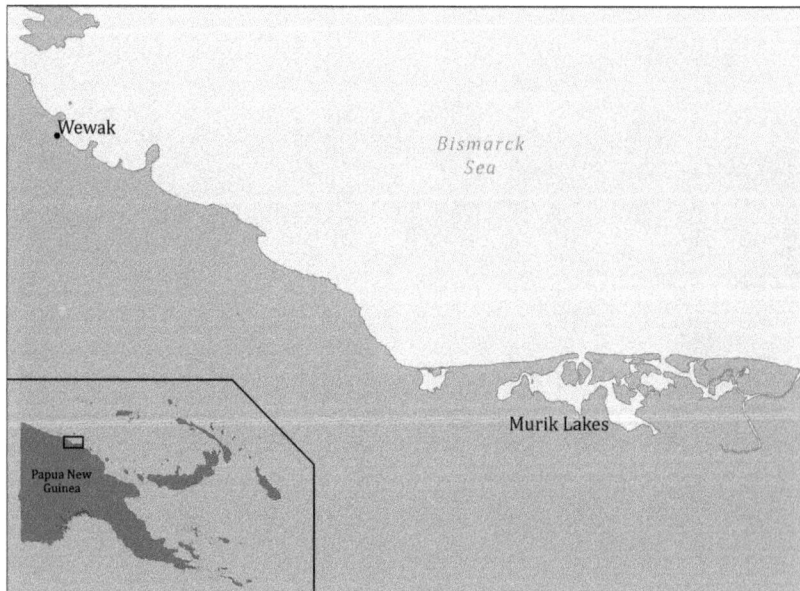

Figure 5. Wewak town and the Murik Lakes
Source: David Lipset, U-Spatial, University of Minnesota

---

1   My first visit to Darapap village took place in 1981 and my last visit was in 2012. This was my 11th visit.

Now, aside from my fellow passenger's uncertainty about what it meant, the successful coordination of urban–rural time and space upon which the preceding scene had been contingent was based on mobile communication, the moral economy of which I propose to explore in this chapter (Ling and Campbell 2008). Although the Murik communities, and the Lower Sepik region more generally, were not then served by Digicel, villagers knew of certain spots where rather unreliable signals could be received and sent from handsets. Signal strength, although never very strong, was known to fluctuate at certain times of the day. In Darapap village, signal reception is located at a particular stretch of beach for an hour or two at and after dawn, or in a doorway of a house.

In this instance, the village councillor, who was then in town, had alerted folks back in Darapap via a handset dangling in the aforementioned doorway that I was in Wewak and was intending to make my way yet again to the village, arriving around midday the following day. I am viewed there with some equivocation, I must say, partly as adoptive kinsman and a 'Darapap son' and, inevitably, as an elite dignitary. But, apparently, there was some sentiment among men in my age cohort to honour my arrival with a welcoming ceremony (which had never been staged for any other of my returns).

In the event, my fellow passenger's irritation had apparently arisen from the moral economy that the performers took for granted. The dance was a gesture of hospitality for which he said he felt unprepared and therefore humiliated. Having not been told about it, he had not prepared anything for me to reciprocate the dancers, the conventional counter-gift being an 18-kilogram bag of rice, a case of tinned fish and a carton or two of beer, which I should have deposited in the middle of the dancers' circle.

Mobile phone communication of the sort I found going on between town and the Murik villages may be positioned in between digital piracy and Grameen-like, village phone sharing (Cohen 2001).[2] It is not part of an illegitimate economy because its relationship to the corporate

---

2   The Grameen model refers to a payphone system in rural Bangladesh under which micro-entrepreneurs, who are mainly women, borrow money for a special landline configured for multiple-user accounts and rural access via powerful antennas. The entrepreneur purchases minutes in bulk, which she resells to customers in her village. She gets a livelihood and her village gets connectivity. The approach has been assessed not only in terms of business efficiency, scalability and sustainability, but also in terms of gender empowerment and social transformation (Aminuzzaman 2002; Aminuzzaman, Baldersheim and Jamil 2003; Bayes 2001).

infrastructure is not illegal. In order to make calls from serendipitous spots where a signal reception takes place, Digicel credit units must be used. Moreover, the reception of the calls is not the result of any kind of bypass or siphoned-off solution that is accessed outside legal structures. Legal rights are not infringed. Civil society is not contested via the use of unauthorised media connections. Mobile phone use outside the network is not an illegitimate double, like Nigerian bootleg copies of feature-length, Hollywood films (Larkin 2004). Rather, its similarity to piracy consists in the way in which the network has exceeded its boundaries, the degradation and technological irregularity of its quality outside of those boundaries and, lastly, in the avid interest of a population living there to link their peripheral community to kin in the Melanesian modernity emerging in Papua New Guinea. So, rather than corporate piracy, the main purpose of the mobile phone sharing that I found on this periphery was moral. It was collective. It was meant to integrate dispersed kin across time and space in the nation-state.

Mobile Telephones in Urban Melanesia

Since the 2005 decision of the government of the then prime minister Michael Somare to deregulate competition among information and communication technology (ICT) providers in the country, and the entrance of Digicel into the mobile phone market two years later (see Stanley 2008: 26), attitudes about and expectations of the impact of mobile phones in Papua New Guinea have been lively, to say the least (Temple 2011: 59). Government officials hoped that modernity in Papua New Guinea was now only a phone call away. When introducing the new PNG National Information and Communications Technology policy, then minister for public enterprise, information and development cooperation, Hon. Arthur Somare, saw ICT and was consumed with longing. ICT, he declared, had done nothing less than:

> transform … the rest of the world, providing opportunities for businesses, opening access to the global marketplace, delivering a wealth of information, enhancing social interaction, enabling greater community participation and bridging the digital divide that may develop as a consequence of the advancement of technology over time. (Somare 2007)

Grey literature from The Australian National University predicted that ICT in Papua New Guinea would increase information flows and make government more transparent by allowing bloggers to expose corruption and demand accountability (Logan 2012). Mobile phones would also increase women's access to education. In a working paper on money transfers in the Asia-Pacific region, Singh and Nadarajah (2011) predicted that 'mobile money' – transferred via mobile phones – would facilitate payments for cash crops, such as coffee, and benefit Papua New Guinea's remittance economy. Notably, major banks (Tainda 2011; Bank of the South Pacific 2011) began targeting rural customers in the country to whom they referred as the 'unbanked'. One bank partnered with mobile communication service providers and the sole power company in the country, the state-owned PNG Power Ltd, to enable consumers to purchase electricity using mobile phones (Robby 2012).

Research also extolled its contributions to healthcare delivery in the new modernity. Au (2009) studied 'telehealth' in several provincial hospitals and found that young male physicians and paramedics used ICT resources to better their knowledge and skills. Yamo (2013) did a subsequent project on the impact of mobile phone use in healthcare services in the Western Highlands. He interviewed a health officer who described the delivery of a twin baby in a breech position via instructions given over a phone, which avoided the necessity of a caesarean section. Suwamaru (2014) found the following principle uses of mobile phones in urban Papua New Guinea: for business purposes; as an Internet portal; to maintain contact with, enable remittances to and exchange phone credit with kin and friends; and, lastly, to contact healthcare services and schools. In particular, many people praised the 'instant contact' that ICT allowed with kin.

Generally speaking, official attitudes and academic research acclaimed mobile phones for facilitating rational communicative action – to evoke Habermas's notion of modernity (1984). But purposes that defy rational incorporation have also been noticed. Simmel (1978) and others (Goffman 1967; Berger 1979, 1992; Giddens 1990, 1991; Seligman 1997) argue that modern pluralism and anonymity permit the self to initiate autonomous relationships with the other; for example, in courtship and friendship. At the same time, such independence raises a problem of trust in modern relationships, with trust referring to the moral expectation of reliability and honesty in dialogue. Now it is true that I, as well as others, have argued that forms of dissimulation (Lipset 2013), or what Andersen has called 'lies and trickery' (2013), preceded modernity in PNG societies (see also Kulick 1992). But mobile telephones seem to have enabled the development of peculiar forms of moral economy. I am referring to *gesfaia*, which is Tok Pisin for random phone calls made by anonymous callers – strangers who may be either men or women, but are mostly made by men. These calls are made to unknown people with whom lengthy conversations are then conducted, 'phone friendships' may then evolve in which gifts of phone credit are transferred (Andersen 2013; Jorgensen n.d.).[3] Andersen's informants, who were young women in a nursing school in a Highlands town in 2011–12, took pleasure from such conversations by answering their anonymous callers, whom they presumed were lying about their marital status, with no less made-up stories about their own status. In other words, the *gesfaia* calls enabled by mobile telephones disembedded voices of the young in particular, from moral time and space. They suggest the rise of a decidedly Melanesian form of individualism, one that is free of social constraints imposed by kin and foreground the ambiguity of contemporary urban life.[4] Mobile phone calls refer to the self rather than to place and its moral community. They create a version of privacy in which the self prefers morally ambiguous relationships with the other to social isolation.

I also want to draw attention to a couple of other intriguing and relevant uses of mobile phones by youth in the capital of nearby Vanuatu. On the one hand, inboxes and outgoing call logs on mobile phones have become a record of licit and illicit intimate moral interaction – thus becoming a legible kind of gossip. On the other hand, however, young people are known to deliberately turn off their phones in order to 'disconnect' from the other. They do not seek solitude or privacy. They do not seek to preserve a communicative space for an individuated self, but are rather motivated by a desire to 'manipulate their partners' emotions, hoping to rekindle [their] … feelings of love, need or attachment' (Kraemer 2017: 50; cf. Archambault 2013).[5] In the world they inhabit, where the love, care and support of kin is often lacking, the agency to turn off the new technology is not meant to give rise to independence from society, it is rather symptomatic of the precariousness and *anomie* of urban life. They try to maintain personal relationships above all else, for all of their social and material values, as they may be.

---

3   Andersen (2013) also claims that Digicel representatives were among the first to engage in *gesfaia* calls to young women giving them phone credits to encourage consumer uptake.
4   Similarly, Kraemer reports that 'young people often' ask strangers whether they 'have a mobile phone' by way of seeking their permission to call or text them at some point in the future (2015).
5   Women also expressed concerns that mobile phones were used for sorcery and/or witchcraft, fears that led many of them to turn their phones off at night (Kraemer 2013: 178).

Mobile Telephones in the Rural Melanesia

Thus far, I have discussed the ethnography of mobile phone technology in Melanesian cities, where the infrastructure of the postcolonial state is evident, if rather unpredictably so. In many rural villages, as I pointed out at the beginning of this chapter, not only is infrastructure, such as electricity and water, lacking, so is signal reception. What is not lacking is a keen desire for connectivity. Writing of a remote people living along the Upper Frieda River, who were on the verge of hosting the construction of a major gold mine, Sullivan reported that they 'explicitly asked to bring in towers' (2010: 3). Likewise, Watson (2011) found that, after Digicel's mobile coverage rollout in the first years after 2007, basic mobile handsets sold out in Madang Province to excited rural people who barely knew how to operate one. In no-service areas, rural villagers search for or unexpectedly come upon random spots where signal reception, although weak, may be accessed – in trees, for example (Telban and Vávrová 2014). Who do they want to call?

The few reports we do have about rural mobile phone use emphasise its embeddedness in social structure and cosmology and, willy nilly, its relationship to modernity. In a Vanuatu-based case, we hear of an overriding wish to stay in touch with kin (Lindstrom 2011). In another, mobile phones are said to aid chiefly claims in ritual status competitions. 'Mobile phones are routinely praised by … men for facilitating the complex exchanges of pigs and other forms of wealth that take place in male grade-taking ceremonies' (Taylor 2015: 8–9). Likewise, in the Southern Highlands province of Papua New Guinea, Macdonald has discussed contemporary tribal warfare among Kewa people that had been suppressed by the Australian administration in the mid-20th century but re-emerged in the 1990s in a more deadly, yet recognisable, form. Now armed with powerful modern weapons and new motives, warriors sent texts to coordinate fighting, the phone credit for which was supplied to them by middle-class Kewa elites (Macdonald and Kirami 2016: 113–14).

Two practices among a remote Upper Sepik River community shed more light on the embeddedness of mobile phone use in social structure and cosmology. Lacking a signal in the village, Ambonwari people turn on phones at the end of a 10-hour trip downriver to the district seat, where there is reception. They call kin in town with a heads up about their imminent arrival – and to ask for phone credit. Once in town, they turn speakers off and leave them off, except in the company of kin. They also refrain from giving out phone numbers. When their numbers become too well known and they start receiving 'suspicious ... calls, they ... buy a new SIM card and get a new number' (Telban and Vávrová 2014: 228). Certain phone numbers, moreover, have to be kept absolutely secret because they are understood to be a way to speak to the dead – deceased sons, in particular – who could make bank deposits on behalf of parents and give them access to 'goods and ways of life which are typical of otherwise hidden realms inhabited by the dead and white people' (Telban and Vávrová 2014: 237). In other words, in their transition to town, Ambonwari mobile phones enter modern spaces of anonymous sociality and resources – modern spaces of which they are both distrustful yet which inspire desire. And this is the point I foreground in this chapter: how the new connectivity is informed by the moral economy that predates it and thus constitutes an emergent Melanesian modernity. As described in my introduction, I encountered a relatively clear-cut, although rather differently toned, expression of this kind of vernacular modernity in 2014, upon my return to the Murik villages.

Phone-sharing in a Lower Sepik Village

In addition to being a zone of no economic development, Papua New Guinea provides no infrastructure to the Murik Lakes, other than the occasional outboard motor and boat, money and primary education. There are plenty of commodities in circulation around the villages, however, ranging from bags of rice, fishhooks, second-hand clothing to radios, generators and wristwatches, amongst much else. But the Murik small-scale fishery has struggled to find its market in recent years, particularly in light of increasing costs of fuel. The moral economy combines reciprocities among kin and mini-market exchanges; also among kin, a combination of local and modern values and norms that, by contemporary standards, might equate to a kind of poverty. People go on living in bush material houses, doing subsistence fishing on a daily basis and buying processed sago flour and garden produce from hereditary trading partners – purchases that are also meant for subsistence consumption. There is no electricity, no water and no sewage. The Digicel coverage map shows a gap in the service it provides: its network extends to perhaps 60 per cent of the province, but excludes both the Murik Lakes and the rest of the Lower Sepik. People have radios and, therefore, they do not entirely live off the grid. They listen to the state-owned National Broadcasting Corporation, which broadcasts from Port Moresby and from its regional satellite station in Wewak, the capital of the East Sepik Province.

Figures 6–8. Some spots where signal reception was possible in the Murik Lakes

Source: Photos by David Lipset, 2010–12

But villagers have a strong desire for 'perpetual connectivity' with kin. Indeed, the force of what might be called their collective id in this regard is inexhaustible. To access the network, I have seen ladders propped up against trees, platforms built in mangrove trees, as well as people standing on particular spots of beaches at odd times of the day (see figures 6–8). In each of these places, mobile signals are said to be momentarily, but unpredictably, accessible and individuals with power in the batteries of their handsets and credit units to burn, repair there and try to call kin in Wewak town or elsewhere in the country. In 2014, a handset hung by a line in an interior doorway of a small house in Darapap village. The venue was the private residence of a young married couple, Venis and Erik, who lived there with two children. During early mornings, late afternoons and at night, a signal reached this distinctively van Gennepian spot (1960), and incoming calls were received from kin living in Wewak town. Here, in the space in between a kitchen and a bedroom, was a portal to and from urban modernity (see Figure 9). What was the relationship of the device to village society? What concept of 'public' did it constitute in a kinship-based community? And what impact did it have on rural time and space?

While Venis admired how the dangling phone enabled communication over long distances (Ling and Yttri 2002), she also pointed out that, on occasion, it gave rise to confusion over physical and social space. When she went to town, and brought her phone with her, sometimes people called expecting to have reached the village. 'Should somebody ring while I am in town [with my phone], I will say, "Sorry, I am not in the village", and they don't call back.' In other words, the dangling phone does not enable the kind of anonymous calls to which I referred above. It is a link to another kind of modernity.

When I asked how she and her husband managed the only working phone in Darapap, Venis phrased her answer in terms of the gift – for example, in terms of a moral economy of sentiment and generalised reciprocity – rather than in terms of a commoditised service (Sahlins 1974). She viewed phone access as the provision of a collective service. She and her husband had turned the dangling handset, whose battery she recharged with a solar panel,[6] into a phone booth for which she served as intermediary. 'Whatever kind of call comes to the house,' she allowed, 'I take a message to whomever it is meant for.' She elaborated a little about what she meant

---

6   The provincial government started a fisheries project a few years earlier for which it had erected small freezer houses with solar panels and batteries in each of the Murik villages. The project never got beyond this point, however, and the equipment was eventually dismantled by villagers and the solar panels were put in use as their 'owners' saw fit.

by the phrase 'whatever kind of call': they might be angry, or they might be open-ended. But they were all personal in nature. That is, giving voice as they did to kin living in a particularistic social order, they were made in a space that was both public yet private.

Figure 9. A mobile phone hanging by a doorway in Darapap village
Source: Photo by David Lipset, 2014

She did not sell flexcards, she went on, and did not charge people to use their phone. 'It is just free,' she said. 'If somebody gets upset or concerned about something here in the village, I … help them with my own phone credits. I … help them call.' Some people, she added, buy their own SIM cards and have their own credits and go ahead and use her phone, again at no charge. But she and her husband ask 'people in town to … transfer [credit] units to … [their phone] in return for helping them ring their kin'. In addition, the schoolteachers use the phone and give them 'a unit or two in return'.

I asked Venis to keep a list of incoming and outgoing calls over a three-day period and then discussed the calls with her. Several found the intended recipient unavailable for one reason or another. Only two outgoing calls were made during this interval: a brother made one to his sister in town to confirm that the boat taking me to the village had indeed departed and the other was made by a husband to his wife to confirm that he had given two bags of rice to her kin in the village. Most incoming calls also expressed ordinary concerns and interests of kin for one another. One of Venis' sons called to ask how she and the rest of the family were doing. A female in-law of one of the schoolteachers called to report that one of her 'aunties' was ill. A call came from the wife of my research assistant to check whether he had given some bags of rice she had sent with him to her kin living in the family fishing camp in the lagoons at a distance from the village (Lipset 2014). A husband called regarding his wife's health and asked Venis to check on her wellbeing. A young woman called to inquire about the whereabouts of her boat and outboard motor and got angry when Venis told her that it was not in the village but was docked at a Lower Sepik River village. Another call came from a relative of Venis to say that a boat on its way to Madang to deliver betel nuts to Highlands buyers would stop off to pick up her husband and take him to market fish there. A woman called asking whether her husband, who had also left the village on a betel nut boat, had returned. My research assistant's wife texted him to call her back because she wanted to confirm that the boat to pick us up and take us back to town would be arriving shortly. He called her back but the connection was weak, which Venis blamed on the high tide. She called back later, but he was no longer near the phone. Venis answered the phone and told her, 'The network is weak. Excuse us!' The battery then died and had to be recharged, which took the rest of the available daylight.

In addition to these incoming calls, the deaths of two young men, Wesley and Frankie, which had recently taken place in Wewak town, was the topic of several calls. Three calls, again from my research assistant's wife, the second from the son of a father's brother of one of the two young men, and the third from a granddaughter of another father's brother, all confirmed the deaths. One of the deaths, it was reported, had not been expected. While Frankie, it was explained, had been unable to keep food down for a period of time, Wesley's death apparently happened out of the blue. A son of Venis called to update his mother about plans for the transport of the two bodies back to the village. They were to be returned for separate funerary rites and burials (Lipset 2016). A brother of Wesley called to dispatch somebody to an inland village where kin lived who would give them sago flour for the funerary feasts. Another brother then called to say that a coffin was being finished and that the body was expected to be taken back to the village the following day. The councillor called twice more. He sent messages to his brother that the latter man's wife would boat down the coast that night and would explain how they wanted to organise the funeral; they must wait before opening the 'death house' to mourners to start their all-night vigil. Instead, the elder brother of Frankie should paddle to the neighbouring village of Big Murik, where the corpse would first 'overnight' so that family there might grieve.

Clearly, this three-day 'sample' of incoming calls does not exhaust the uses to which the one phone in Darapap village is put (much less elsewhere in rural Papua New Guinea). But I think that several generalised points about the relationship of mobile telephony to rural society may reliably be read in these data: 1) mobile phones expand the physical, and, more importantly, the moral space of kinship; 2) they enable space–time compression; for example, a coordination and simultaneity over distances for individual actors in the moral community; 3) the dangling handset is private property but it is not a private device – it does not create privacy. It is not used for duplicitous purposes of an immoral self in an immoral network. While it links this rural village to town and thus, by implication, to the nation-state, its predominant use is not personal but social, inclusive and situational in a decidedly ordinary sense; which is to say, 4) mobile phone sharing in this venue is minimally linked to modernity, to the marketplace, anonymous modernity and ego-centred interests and desires. Rather, it expands space and time informed by normative, kinship-based values and the moral economy of the gift.[7]

Conclusion: Mobile Phones and Melanesian Modernity

Regimes of capital, broadly defined, were once seen to depend on state-based infrastructure to provide the preconditions for the circulation of commodities, energy, information and so forth. Today, in an increasingly neoliberal political environment, state-based infrastructure no longer integrates diverse places into temporal and spatial relationships, but is supplemented by private, corporate institutions. Infrastructure, however, is created that includes elites and townspeople while it marginalises rural folk. The present case, signified by a mobile phone hanging in a doorway, is clearly one of the latter communities, one that barely straddles the digital divide. Corporate infrastructure, the Digicel network, has not integrated nor excluded Darapap village in any unitary way. Deeply contradictory, the communicative world of Digicel, at least from the viewpoint of this remote village, adheres neither to logics of legitimate micro-enterprise nor piracy. It enacts no narrative of utopian transformation or defiance. It has rather produced a paradoxical partnership with modernity that is both inside and outside, connected yet detached.

People, ideas and capital crisscross nations, regions and the globe at large in this day and age. I think that the mobile phone might be viewed as emblematic of anytime, anywhere communication that brings the ubiquitous quality of movement characteristic of this historical moment into experience. The mobile phone does indeed make the flow and friction of information across spatio-temporal boundaries concrete in everyday life, but this does not necessarily mean that the devices transform existing places, temporalities or moral identities beyond recognition. To the extent that the new order may involve the enlargement of social scale,

---

7   De Souza e Silva (2007) reported that, in Brazilian *favelas* (slums), where infrastructure was also lacking, similar informal appropriations of the new technology have occurred wherein people put shared mobile phones to work for whole community use (see also Castells et al. 2007).

however, in this chapter I have shown that it may also be associated with the maintenance of moral networks linking rural communities and places with their urban diasporas.[8]

References

Aminuzzaman, S. 2002. Cellular Phones in Rural Bangladesh: A Study of the Village Pay Phone of Grameen Bank. In *Electronic Commerce for Development*. A. Goldstein & D. O'Connor, eds. Paris: OECD Development Centre Studies, pp. 161–78. doi.org/10.1080/0958493032000175879

Aminuzzaman, S., H. Baldersheim & I. Jamil 2003. Talking Back! Empowerment and Mobile Phones in Rural Bangladesh: A Study of the Village Phone Scheme of Grameen Bank. *Contemporary South Asia*, 12(3): 327–48. doi.org/10.1080/0958493032000175879

Andersen, B. 2013. Tricks, Lies, and Mobile Phones: 'Phone Friend' Stories in Papua New Guinea. *Culture, Theory and Critique*, 54(3): 318–34. doi.org/10.1080/14735784.2013.811886

Archambault, J.S. 2009. Being Cool or Being Good: Researching Mobile Phones in Mozambique. *Anthropology Matters*, 11(2): 1–9.

—— 2013. Cruising Through Uncertainty: Cell Phones and the Politics of Display and Disguise in Inhambane, Mozambique. *American Ethnologist*, 40(1): 88–101. doi.org/10.1111/amet.12007

Au, L. 2009. Assessing the Potential Needs for Telehealth in Papua New Guinea (PNG). MA Thesis. University of Canterbury, New Zealand.

Bank of the South Pacific 2011. Mobile Banking. www.bsp.com.pg/Personal/Ways-to-Bank/Mobile-Banking/Mobile-Banking.aspx (accessed 10 April 2017).

---

8   Work in Jamaica (Horst and Miller 2006), migrants in China (Cartier, Castells and Qiu 2005) and in South Africa (Skuse and Cousins 2007) all illustrate linkages between urban and rural communities.

Bayes, A. 2001. Infrastructure and Rural Development: Insights from a Grameen Bank Village Phone Initiative in Bangladesh. *Agricultural Economics*, 25(2–3): 261–72. doi.org/10.1111/j.1574-0862.2001.tb00206.x

Berger, P. 1979. *The Heretical Imperative: Contemporary Possibilities of Religious Affiliation*. Garden City, NY: Anchor Books.

—— 1992. *A Far Glory: The Quest for Faith in an Age of Credulity*. New York: Anchor Books.

Cartier, C., M. Castells & J. Linchuan Qiu 2005. The Information Haveless: Inequality, Mobility, and Translocal Networks in Chinese Cities. *Studies in Comparative International Development*, 40(2): 9–34. doi.org/10.1007/BF02686292

Castells, M., M. Fernandez-Ardeveol, J. Qui Linchan & A. Sey 2007. *Mobile Communication and Society: A Global Perspective*. Cambridge, MA: MIT Press.

Cohen, N. 2001. What Works: Grameen Telecom's Village Phones. World Resources Institute: Digital Dividend. pdf.wri.org/dd_grameen.pdf (accessed 10 April 2018).

De Souza e Silva, A. 2007. Cell Phones and Places: The Use of Mobile Technologies in Brazil. In *Societies and Cities in the Age of Instant Access*. H.J. Miller, ed. New York: Springer, pp. 295–310.

Giddens, A. 1990. *The Consequences of Modernity*. Cambridge: Polity.

—— 1991. *Modernity and Self-identity: Self and Society in the Late Modern Age*. Stanford University Press.

Goffman, E. 1967. *Interaction Ritual: Essays on Face-to-Face Behavior*. Garden City, NY: Anchor Books.

Habermas, J. 1984. *The Theory of Communicative Action*, vol. 1, *Reason and the Rationalization of Society*. Thomas McCarthy, trans. Boston: Beacon Press.

Horst, H. & D. Miller 2006. *The Cell Phone: An Anthropology of Communication*. Oxford: Berg.

Jorgensen, D. n.d. Domi and 'the mobile phone': Apolcaypse and Salvation in PNG's Wireless Network. Ms draft, with author's permission.

Kraemer, D. 2013. Planting Roots, Making Place: An Ethnography of Young Men in Port Vila, Vanuatu. PhD Thesis. Department of Social Anthropology, London School of Economics.

—— 2015. 'Do You Have a Mobile?' Mobile Phone Practices and the Refashioning of Social Relationships in Port Vila Town. *The Australian Journal of Anthropology (TAJA)*, 28(1): 39–55. doi.org/10.1111/taja. 12165

Kulick, D. 1992. *Language Shift and Cultural Reproduction: Socialization, Self, and Syncretism in a Papua New Guinea Village*. Cambridge: Cambridge University Press.

Larkin, B. 2004. Degraded Images, Distorted Sounds: Nigerian Video and the Infrastructure of Piracy. *Public Culture*, 16(2): 289–314.

Lindstrom, L. 2011. Urbane Tannese: Local Perspectives on Settlement Life in Port Vila. *Journale de la Société des Océanistes*, 133: 255–66. doi.org/10.4000/jso.6461

Ling, R. & S.W. Campbell 2008. *The Reconstruction of Space and Time: Mobile Communication Practices*. Piscataway, NJ: Transaction.

Ling, R. & B. Yttri, 2002. *Perpetual Contact: Mobile Communication, Private Talk, Public Performance*. Cambridge: Cambridge University Press.

Lipset, D. 2009. A Melanesian Pygmalion: Masculine Creativity and Symbolic Castration in a Postcolonial Backwater. *Ethos*, 37(1): 50–77. doi.org/10.1111/j.1548-1352.2009.01031.x

—— 2013. Mobail: Moral Ambivalence and the Domestication of Mobile Telephones in Peri-Urban Papua New Guinea. Culture. *Theory and Critique*, 54: 335–54. doi.org/10.1080/14735784.2013.826501

—— 2014. Place in the Anthropocene: A Mangrove Lagoon in Papua New Guinea in the Time of Rising Sea-levels. *HAU: Journal of Ethnographic Theory*, 4(3): 215–43. doi.org/10.14318/hau4.3.014

—— 2016. The Knotted Person: Death, the Bad Breast, and Melanesian Modernity among the Murik, Papua New Guinea. In *Mortuary Dialogues: Death Ritual and the Reproduction of Moral Community in Pacific Modernities*. David Lipset & Eric K. Silverman, eds. New York: Berghahn, pp. 81–109.

Logan, S. 2012. Rausim! Digital Politics in Papua New Guinea. SSGM Discussion Paper 2012/9. ssgm.bellschool.anu.edu.au/sites/default/files/publications/attachments/2015-12/2012_9_0.pdf (accessed 14 April 2018).

Macdonald, F. & J. Kirami 2016. Women, Mobile Phones and M16s: Contemporary New Guinea Highlands Warfare. *The Australian Journal of Anthropology*, 28(1): 104–19. doi.org/10.1111/taja.12175

Robby, B. 2012. BSP Links up with Digicel, Citifon to Enable Customers to Buy EasiPay Units. *National*, 3 May.

Sahlins, M. 1974. *Stone Age Economics*. Piscatway, NJ: Transaction.

Seligman, A. 1997. *The Problem of Trust*. Princeton: Princeton University Press.

Simmel, G. 1978. *The Philosophy of Money*. Tom Bottomore and David Frisby, trans. London: Routledge & Kegan Paul.

Singh, S. & Y. Nadarajah 2011. School Fees, Beer and 'Meri': Gender, Cash and the Mobile in the Morobe Province of Papua New Guinea. Institute for Money, Technology & Financial Inclusion, Working Paper 2011-3.

Skuse, A. & T. Cousins 2007. *Managing Distance: The Social Dynamics of Rural Telecommunications Access and Use in the Eastern Cape, South Africa*. Information Society Research Group. web.archive.org/web/20060725052535/www.isrg.info/ISRGWorkingPaper1.pdf (accessed 9 April 2018).

Somare, A. 2007. Speech on the occasion of the launch of the National Information and Communications Technology Policy, Government of Papua New Guinea, Waigani.

Stanley, L. 2008. The Development of Information and Communication Technology Law and Policy in Papua New Guinea. *Pacific Economic Bulletin*, 23(1): 16–28.

Sullivan, N. 2010. *Fieldwork Report in Support of an Environmental and Social Management Framework*. Madang, PNG: Nancy Sullivan Associates.

Suwamaru, J.K. 2014. Personal Experiences with Mixed Methods Research in Papua New Guinea. *Contemporary PNG Studies: DWU Research Journal*, 21: 73–86.

Tainda, L. 2011. Microbank and Digicel Partner. *National*, 22 November.

Taylor, J.P. 2015. Drinking Money and Pulling Women: Mobile Phone Talk, Gender, and Agency in Vanuatu. *Anthropological Forum*, 26(1): 1–16. doi.org/10.1080/00664677.2015.1071238

Telban, B. & D. Vávrová 2014. Ringing the Living and the Dead: Mobile Phones in a Sepik Society. *The Australian Journal of Anthropology*, 25: 223–38. doi.org/10.1111/taja.12090

Temple, O. 2011. Tok Ples in Texting and Social Networking: PNG 2010. *Language and Linguistics in Melanesia: Journal of the Linguistic Society of Papua New Guinea*, 29: 54–64.

van Gennep, A. 1960. *The Rites of Passage*. A. Vicedom, trans. Chicago: University of Chicago Press.

Watson, A.H.A. 2011. The Mobile Phone: The New Communication Drum of Papua New Guinea. PhD Thesis. Queensland University of Technology.

Yamo, H. 2013. Mobile Phones in Rural Papua New Guinea: A Transformation of Health Communication and Delivery Services in Western Highland Province. MA Thesis. School of Communication Studies, Auckland University of Technology, New Zealand.

2

HIV, Phone Friends and Affective Technology in Papua New Guinea

Holly Wardlow

In the field of global health, and especially in the field of HIV treatment and care, mobile phones are primarily perceived as a technology for improving adherence to antiretroviral medications and for preventing 'loss to follow up' – that is, preventing patients' unexplained failure to return for ongoing clinical care. If, for example, one does a Google Scholar search on 'mobile phones and HIV', a typical result is 'Mobile Phone Technologies Improve Adherence to Antiretroviral Treatment in Resource-Poor Settings: A Randomized Controlled Trial of Text Message Reminders' (Pop-Eleches et al. 2011). Motivating this kind of research is the concern that, although the expanding provision of antiretroviral therapy (ART) has dramatically decreased HIV-related deaths and illness around the world, irregular or episodic treatment is the major cause of both treatment failure and the development of drug resistance. Thus, the hope is that mobile phones can improve adherence through having healthcare workers or family members text patients to remind them of upcoming or missed appointments, or to take their ART. Mobile phones are imagined instrumentally as a means for improving health outcomes through the enhanced surveillance and disciplining of patients' medication practices.

In Tari, Papua New Guinea, women living with HIV speak often of the importance of mobile phones in their lives, but never as tools for improving medication adherence. Rather, in the face of community stigma or family abandonment, they use mobile phones to collect 'phone friends' who may provide emotional comfort, mental diversion, romantic escapism and even material support. Phone friends can be thought of as intimate strangers – they are people met and known only by mobile phone (Andersen 2013; Jorgensen 2014). Sometimes these connections are made through friends of friends, but most often they are made through calling or accepting calls from unknown numbers. In this chapter, I first provide an overview of mobile phone use in Tari, focusing on the phenomenon of phone friends. I then explain why the sociality of women living with HIV is often severely circumscribed because of both family rejection and their own strategies for self care. I conclude with a discussion of how these women use mobile phones to achieve a more expansive and emotionally fulfilling sociality, showing that, in contrast to the domain of public health, which imagines the mobile phone as an instrumental technology, mobile phones are better understood as an 'affective technology' (Lasen 2004; Silva 2012) – a communication technology that not only mediates the expression and experience of emotion, but also opens up new forms of emotional intimacy.

The following is based on participant observation over four research periods from 2010–13 and on interviews with 28 HIV-positive women taking ART. With two-thirds of these women in mid to late middle age, the group was, on the whole, older than participants in much of the ethnographic research on either HIV or on mobile phone use; both of these areas of research have tended to focus on youth or people of early to mid-reproductive age. Nine of the women interviewed had no formal education, six had completed two or three grades of primary school, 12 had completed sixth grade or gone a year or two beyond sixth grade, and one had completed high school. Not easily able to write or not in the habit of doing so, most of my female interlocutors were using their mobile phones for calling, not texting, and none were using social media.

Mobile Phones in Tari

When I left Tari in 2006, there were no mobile phones and, indeed, few working landlines; when I returned in 2010, mobile phones were seemingly everywhere. Digicel and Bemobile were competing for customers, and a few of my friends owned two phones, one for each network. Feelings of loyalty to the nation often meant that they were rooting for Bemobile, a Papua New Guinean company, to succeed, but they acknowledged that Digicel offered better coverage, more possibilities for free credit and more affordable phones (Jorgensen, Chapter 3; Foster, Chapter 6). The rapid construction of mobile phone towers in the region, primarily by Digicel, was not, however, matched by an expansion of electrical power, and thus the primary challenge for mobile phone users in the Tari area has been finding ways to keep their phones charged. Tari only began to be incorporated into PNG Power's electrical network in 2012 and, even then, power was confined to the main town and was dependent on a 500 KVA generator set that ran only sporadically.[1] The hospital, the small Porgera Joint Venture Community Relations office, the airport and a few stores had their own generators. Not surprisingly, then, every possible electrical outlet in these places and in government buildings was always being used to charge phones, often with a pile of waiting phones littering floors and counters. People who had no kin or friends with access to an electrical outlet were resourceful in their phone-charging strategies: in 2010, I came upon a young man on a remote, mountain road who had five D batteries lined up inside a bamboo tube with wires connecting the tube to his phone (the trick to charging the phone battery without destroying it, he said, was to use a mix of new and old batteries). By 2012, entrepreneurial young men had bought generators and set up charging stations in the main market that could accommodate 10 to 20 phones at a time, demanding 2 kina for a full charge.

Although I focus on the phenomenon of phone friends in this chapter, my female friends and interviewees knew most of their phone contacts. Women took great pleasure and comfort, for example, in being able to speak to their children who were away at boarding school or to mothers and sisters living elsewhere, often at night before bed, asking what had been eaten for dinner and sharing a prayer or blessing. In a virilocal society where it is the wife who moves to her husband's clan territory upon marriage, it is not surprising that mobile phones have been valued by women primarily for allowing them to maintain intimate contact with close female kin (see also Tacchi, Kitner and Crawford 2012). This is acutely so among the Huli, where a wife is expected to ask her husband's permission to leave the household, and a woman's female kin are often discouraged from visiting (Wardlow 2006). Mobile phones have facilitated a degree of post-marital

---

1   Before 2012, the town relied on a small hydropower station that was built and maintained by missionaries in the 1980s; by the early 2000s, it was in disrepair and not easily fixed.

contact between wives and their natal kin that was never before possible. In other words, mobile phones can overcome not only the spatial distance created by physical geography, but also the socio-spatial distance created through gendered relations of power and rules of propriety.

Most of my interlocutors – male and female – had a number of phone friends in addition to their known contacts. Phone friends, as Andersen observes, may be 'a uniquely PNG response to the communicative possibilities of the mobile phone' (2013: 319). Phone friend calling is the calling of unknown phone numbers with the aim of forming a friendship or romance with a stranger that will likely exist only by phone and may endure for one phone call or for many months. The calling usually takes place at night and seems to be initiated more by men than by women. On seeing an unfamiliar number appear on the screen, women can choose to answer and converse with the caller or not. Of course men, searching for a phone friend, often end up calling other men, or they encounter female call recipients who yell or swear at them. Thus, some male friends showed me how they would make systematic lists of numbers and cross the unsuccessful ones off to ensure that they didn't call those again. And, much as Andersen describes, phone friendships are often heterosexual, romantic liaisons in which women 'let themselves be contacted by unknown men, passing long evenings entertaining them with stories about their daily lives, or with teasing, flirtation, and jokes' (2013: 319).

One of the characteristics of phone friend relationships in Papua New Guinea – and a characteristic that gives phone friendships their titillating pleasure – is that it is understood that either or both parties may be lying about who they are. As Andersen notes, 'there is not normally an expectation of sincerity in mobile phone interactions' (2013: 323). For example, one of my male friends would tell his female phone friends that he was a powerful and wealthy landowner with claims to the ExxonMobil Liquefied Natural Gas (LNG) project area, which was not true. When I expressed scepticism that he would do such a thing, he carried out a phone friend call in front of me using his wealthy landowner persona. He told me that he had once come very close to luring a nursing student to Tari all the way from the coastal town of Madang – she made it to Mt Hagen, he said, before caution got the better of her and she turned around. But, he acknowledged, he didn't really know if she had actually come as far as Mt Hagen – indeed, he didn't really know if she ever got on the bus at all, if she was actually a nursing student, or even if she really lived in Madang. This, he said, was one of the pleasures of phone friends

– he might be getting the better of her and playing her for a fool, or she might be getting the better of him and playing him for a fool, or they might both be assuming feigned identities when speaking to each other, and neither of them ever had to know, and perhaps never *could* know, the truth of the situation.

The LNG project was in its development phase at that time and landowners within the project area were receiving millions of kina in small business development grants, preferential employment and subcontracts; consequently, outsiders were flooding into the area, many of them women hoping to cement relationships with these landowners. So, although unlikely, it was not out of the realm of possibility that a young woman, believing my friend's assertions, might abandon her situation and make the two- to three-day journey to find him. And, even in recollection, he experienced a frisson of both glee and terror that his nursing student might exit the realm of mobile phone fantasy and show up at his door. These elements of secrecy, deception and uncertainty, and the deliberate forestalling of definitive knowledge, have been emphasised in some of the ethnographic literature about mobile phone communication. Archambault, for example, writing of impoverished, peri-urban youth in Mozambique, asserts that mobile phones help to 'preserve and reproduce epistemologies of ignorance or modes of knowing that privilege pretense over truth and transparency' (2013: 89). In her case, young women juggle relationships with a number of different men through deceptions made possible by mobile phone communication in which one cannot actually know where the other person is, what she is doing or with whom.

My friends' stories of their phone friend inventions suggest not only a desire to pretend to be someone else for a moment, or to produce false or uncertain knowledge in another, but also to distance oneself even as one cultivates new intimacies with another. In other words, mobile phones enable hitherto impossible affective relationships with strangers across the nation, but simultaneously allow callers to maintain a safe social distance through deception. Given their tacit mutual knowledge about possible deception, phone friends are not likely to make significant requests of one another, or at least not likely to expect that the other will actually meet such a request: a phone friend might affectionately tease you, listen to your confidences without judgement, or ask you to send phone credit, but he or she will not ask you to assist with a bride wealth payment, plane fare or school fees. And, if he or she does, you can delete and block the number and find a new phone friend who will stay within the bounds of an affective intimacy free from social or material obligation.

Gilbert has similarly observed in her study of young women in Nigeria that 'mobile phones allow young women to stay connected while apart … mobile communication allows young women to cope with the fears of social proximity by *maintaining distance*' (2016: 4, her emphasis). In her case, she found that in a context where young women were competing for jobs, admission to universities and influential social connections, their face-to-face relationships with each other 'were often tainted with acrimony, jealousy, and backstabbing behavior' (2016: 3); the desire for an intimate best friend warred with the knowledge that such intimacy exposed one to possible treachery. Mobile phone relationships with myriad barely known acquaintances offered a middle path, and she found that many of her young female interlocutors preferred to spend their time alone in their rooms, but always in touch with their dozens of mobile phone contacts.

Although my female interlocutors did not suffer the same fears about their face-to-face relations, they did say that phone friends were gratifying because one could simply 'share stories' about one's life and 'they won't ask you for anything'. This was in contrast to relations with kin or boyfriends who would inevitably ask for money, a pig or sex. In other words, what might be most interesting about the phone friend phenomenon are the desires animating this practice, one of which seems to be for a relationship that is not easy to find face-to-face and close to home – the intimate other who is affectively present, but does not enmesh one in the thick obligations of kinship, or the anxiety and anger that these heavier relations can entail. Speaking more specifically about romantic phone friendships, one female friend of mine asserted, 'Phone friends are better than friends who are there – action-action friends. Phone friends are very important. Very important. You can talk to them, laugh with them, they might send you money, and there's no action'. In comparing boyfriends with male phone friends, 'action-action' referred primarily to expectations for sex. Having male phone friends was a way of obtaining male attention, affection and perhaps even some material support (though at that time, one could only send phone credit, not money, via mobile phone), without having to engage in unwanted sex. The more we talked about her life history and 'action-action' male friends, however, the more it became clear that 'action-action' also referred to the larger encompassing domain of male-dominant and often coercive gender relations. When one considers the degree of corporeal violence that PNG women often experience in heterosexual relations (Jolly, Stewart and Brewer 2012), it is not so surprising that some might find disembodied phone friends preferable.

HIV, Stigma and Social Contraction

The phone friend relationships of the HIV-positive women I interviewed differed somewhat from those of other people, and these differences stemmed from their strategies for managing HIV-related precarity, worry and loneliness. People living with HIV/AIDS (PLWHA) in Papua New Guinea have experienced exclusion, isolation and even violence; the condition is associated with both sexual immorality and death, and is sometimes said to be God's punishment for sexual sin (Eves 2008; Hammar 2010; Kelly-Hanku, Aggleton and Shih 2014; Wardlow 2008, 2017). In Tari, AIDS is also associated with an abject and frightening body. As one woman said about the day she was diagnosed:

> Some of the hospital staff said that it was too late, that there was nothing they could do for me, that my kin should wait for me to die and then bury me at the edge of the family cemetery. (At the edge?) Yes, lots of people think you need to bury people who die of AIDS in the wilderness or at the very margin of a family property, far away from everyone. (Why? Because they are scared of getting HIV?) Well, some are afraid of this, but many are afraid that AIDS will make the body rot in a dangerous way. Or they are afraid to look at the face of someone who died of AIDS because they are often so thin, so wasted. They are scared to look at them, afraid to smell them, afraid even to look at their graves – afraid that their spirits will be shocked if they behold them.

The availability of free ART in Tari since 2007, along with widespread AIDS education, has dramatically changed how PLWHA are treated by others, and 19 of the 28 women I interviewed spoke of being supported and nurtured by their families. Emblematic of this social acceptance was the common assertion that others were willing to 'finish my can of Coca Cola' – in other words, willing to put their lips where the woman's lips had been and to consume a beverage that might contain some of her saliva. Some of the women I interviewed, however, had been treated brutally by their families. One young mother, for example, was homeless: upon learning of her diagnosis, her family violently evicted her (she showed me the machete scars on her shoulders and back), and she and her two young children often slept under a tarp in sweet potato fields. Another woman, who had lived in her own small house on her brother's land, had this house burned down with all her clothing and bedding inside, and she likewise was evicted. Less brutally, many families permitted their HIV-positive family member to remain in the extended family compound, but they were expected to eat and sleep in their own small house, and sometimes expected to provision and cook for themselves, which was occasionally impossible if they had health setbacks. In other words, they were not driven off, but they were made to feel that they were a danger to the household.

Consequently, many of the women were in economically precarious situations and had found their social lives far more circumscribed than before. They were not necessarily socially isolated or excluded – some continued to sell produce at market, for example, and many regularly attended church – but they had experienced a contraction in their social networks. There were a number of different reasons for this social contraction. One young woman, for example, had been raped when a gang held up her PMV (public motor vehicle) at gunpoint, and she had become fearful about leaving her household. Other women spoke of being worried about running into kin who might snub or shame them in public. Some worried about accepting others' generosity and being unable to reciprocate because of a reduced ability to do agricultural work, and so they chose to narrow their circle of relations. A common fear was that they might encounter people who would make them angry or distressed by insulting them or reminding them of a past altercation or unresolved debt. During their HIV counselling, they had learned that they should avoid both worry and anger because these two emotions could 'wake up the virus' or give it the strength to 'break the fence' that ART had built around it. Many of them had known people who had been taking ART and died anyway, and they often attributed these deaths to excessive worrying or anger – either the emotions themselves, especially worry, had killed their bearers, or the emotions had roused and disturbed the virus, so that it was able to wreak havoc unchecked. When the women themselves developed rashes, coughs or a persistent fever, they would often attribute these symptoms to a lack of vigilance about controlling negative emotions. One strategy for avoiding dangerous feelings was to limit one's exposure to the social situations that might provoke them. Much like Gilbert's interlocutors who stayed in their rooms in order to avoid the possible treacheries that face-to-face intimacies could entail, some of the women I interviewed stayed within their households in order to avoid the worry or anger that might result from venturing outside.

Phone Friends among Women Living with HIV

All of the HIV-positive women I interviewed who had mobile phones (and approximately one-third of them did not) had phone friends, though the nature of these friendships differed from what Andersen (2013) found in her research with younger, more educated, urban women. For example, Andersen notes that she never heard of women initiating phone friend calls, and she characterises phone friendships as heterosexual and often romantic. In contrast, in addition to having a few male phone friends, many of the women I interviewed had female phone friends, and they themselves had initiated some of these friendships by calling unknown numbers, often at night when feeling abandoned, lonely, worried or desperate about where the next day's food, water or firewood might come from. A couple of them had acquired phone friends not by making or accepting random number calls, but by calling the contacts on used phones (probably stolen) that they had bought cheaply on the street.

I provide one example here. Last Minute Lucy, who looked to be in her early 50s, was sometimes cruelly teased by family members for 'catching AIDS at the last minute', by which they meant at a later stage in life when most people think a woman's sexual activity should be winding down or over. She'd had a disastrous first marriage when she was much younger and, after having been beaten and repeatedly infected with gonorrhoea by her husband, she'd left him and gone through a period of 'passengering around' (having casual sex with multiple partners, sometimes in exchange for money; see Wardlow 2006). She soon tired of this lifestyle, however, and moved into her older brother's household. Dependent on him for land and a home, she had dutifully cooked and cleaned for the household, and she said she hadn't had sex for 40 years. Given my estimation of her age, this did not seem likely, and I understood this statement as signifying both a very long time and, with its possible gesture towards the Jews wandering in the wilderness for 40 years, a kind of self-imposed penance for her earlier unrestrained ways.

> And then, at the last minute, my brother hit me. (Why?) He didn't like the way I'd prepared his dinner. I'd been caring for his household for years, and then one night he lost his temper about having to eat sweet potato all the time, and he hit me. So I left. I decided to passenger around again … I befriended a man, I went home with him, and I caught AIDS.

According to her, after her many years of abstinence, she'd had sex with only this one partner, and he had infected her with HIV. When he began to show symptoms, she insisted that they both be tested and, upon being diagnosed positive, they began a regimen of ART. He did not live his life as the nurses had instructed, however; he took his medication irregularly and continued to drink and sleep around. Lucy left him, saying that he was causing her too much stress and worry, and she was concerned that the negative emotions he provoked in her would render her antiretroviral drugs ineffective. Her older brother refused to take her back in, however, and when I met her in 2012 she was struggling to live on her own on some land her younger brother had made available to her.

When her first husband learned that she was HIV-positive and that her family had refused to take her in, he gave her some money and a used mobile phone. There were still numbers stored in its memory and, one night, in a moment of despair, when she hadn't eaten for two days, did not feel she had the strength to leave her house and was considering suicide, she started calling the numbers. A few of them didn't work, a few people yelled at her and disconnected and then one woman, Angela, answered and agreed to speak with her. It turned out Angela lived in a town about a day's drive away, had attended AIDS awareness workshops and was a devout Christian. She responded to Lucy's despair with compassion, but also reminded her that her situation was part of God's plan. And the next day she sent Lucy some food, second-hand clothing and money with a PMV driver coming to Tari. When I interviewed Lucy she had never met Angela, but they continued to call one another, and Angela still occasionally sent small gifts to her. Angela's generosity towards a complete stranger who had reached out to her by phone in the middle of the night had confirmed for Lucy that her situation was, in fact, somehow part of God's plan: God had used Angela as his instrument (his mobile phone, one might say, or at least his affective technology, restoring her feelings of hope), and she must therefore persevere and not give in to despair.

When I saw Lucy again a year later, she was in anguish because her phone had been stolen; '*tingting bilong mi* fucked up', she said (my thoughts are fucked up). She had come to depend on Angela and her other phone friends – male and female – for emotional nurturance. When she found herself '*tingting plenti*' (thinking a lot), '*wari wari*' (worrying a lot) or '*busy tumas*' (very busy, but it can also mean overly preoccupied with or perseverating about something), she would call them and they would help her calm down or distract her with teasing, chatty discourse. Her phone friends were, in effect, affective regulators, helping her to modulate her anxious thoughts and emotions. And, since such emotions were thought to strengthen HIV and reduce the effectiveness of ART, these friends were, from her perspective, also helping to keep the virus in check. The mobile phone, as affective technology, was thus working simultaneously as a therapeutic technology (but not in the way imagined by global health practitioners), which, to her mind, was equally important as her daily medications. Since she'd never met these friends face-to-face, she had no way of retrieving their numbers or restoring these relationships and she was feeling their loss and her mounting anxiety acutely:

> All those phone friends, in Port Moresby, Mt Hagen, and other places, they would send me credit, and we talked all the time, every night, and now I don't have a phone, and I've lost all those numbers. It's terrible. I can't stop crying about it. I had the same phone for four years, and had so many numbers, so many friends. And now it's all gone.

Lucy's story of her encounter with Angela is probably the most dramatic of the phone friend stories I heard, with its intimation of possible divine intervention into the dark night of the soul, though a number of the women I interviewed similarly asserted that, when they managed to establish a phone friendship with a kind stranger somewhere in Papua New Guinea, they took it as a sign that God was reaching out to them and affirming the value of their continued life. And, like Lucy, many of the women I interviewed said that they were candid with phone friends, male and female, about their HIV-positive status. Since, as discussed above, people often emphasise the possibilities for dissimulation offered by the mobile phone and since most of the women had experienced some kind of rejection or stigma because of their HIV status, this candour might seem unusual. There are a number of reasons for their frank self-exposure. For one, many of them felt both an ethical and self-protective compulsion to disclose their status to their family members, congregations and larger community (Wardlow 2017); I was repeatedly told, 'Everyone knows, everyone in my community knows'. Health workers encourage people living with HIV to disclose their status, at least to spouses and close family members, and many are reluctant to provide ART to patients whose family members don't accompany them to their appointments or who show other signs of hiding their status from close kin. They worry both that these patients do not have adequate social support and that they are not 'reliable' – the euphemistic word they often used to describe patients who might be engaging in unprotected sex or other practices (e.g. gambling, carousing, eating greasy street foods instead of garden produce) that indicate an unwillingness to abide by the expected behaviour of an AIDS patient. Moreover, the women I interviewed knew that, for some community members, not disclosing one's HIV-positive status was equivalent to hiding it and might lead to the suspicion or even accusation that one was *'giaman olsem normal'* (pretending to be normal), engaging in unprotected sex and infecting innocent others. The only way to prevent others' misgivings about one's possible malevolent actions was to disclose assertively and widely. This imperative to disclose might, for some women, be internalised so powerfully that it holds true even with their phone friends, whom they will likely never meet. Perhaps more important, however, is that their phone friends provide them with vital affective sustenance. Since affective regulation and relief is what many of them seek from phone friends, candour about their HIV status is necessary – without it they cannot confide their specific miseries and have them soothed.

Conclusion

Mobile phones can be conceptualised as an affective technology that enters social contexts characterised by specific gendered relations of power and potentially changes them, enabling, for example, closer contact between married women and their natal kin, or romantic heterosexual relations free from coercive or violent corporeality. Further, the pleasures and possibilities of phone friendships in Papua New Guinea may include not only the creation of a fantasy space in which one can, without repercussion or expectation, pretend to be someone else, but also the creation of an affective or therapeutic space in which one can, also without repercussion or expectation, be fully candid about oneself. For HIV-positive women in Tari, mobile phones are clearly more than tools for improving adherence to pharmaceutical and clinical regimes. Women whose networks have been circumscribed by stigma can use them to expand their social worlds, often finding affective sustenance that has been withdrawn following their diagnosis. And, while these relationships are sometimes heterosexual, flirtatious and romantic in nature, they are also sometimes homosocial relations in which women can confide their anxieties and receive compassion, advice and emotional solace.

Acknowledgements

This research was funded by SSHRC Standard Research Grant #331985. I thank Professor Betty Lovai of the University of Papua New Guinea, and the health workers who helped to facilitate this research, especially Martha Parale; and Jethiro Harrison and Ruben Enoch, who were University of Papua New Guinea student interns with me in 2013.

References

Andersen, B. 2013. Tricks, Lies, and Mobile Phones: 'Phone Friend' Stories in Papua New Guinea. *Culture, Theory and Critique*, 54(3): 318–34. doi.org/10.1080/14735784.2013.811886

Archambault, J.S. 2013. Cruising through Uncertainty: Cell Phones and the Politics of Display and Disguise in Inhambane, Mozambique. *American Ethnologist*, 40(1): 88–101. doi.org/10.1111/amet.12007

Eves, R. 2008. Moral Reform and Miraculous Cures: Christian Healing and AIDS in New Ireland, Papua New Guinea. In *Making Sense of AIDS: Culture, Sexuality, and Power in Melanesia*. L. Butt & R. Eves, eds. Honolulu: University of Hawai'i Press, pp. 206–23.

Gilbert, J. 2016. 'They're My Contacts, Not My Friends': Reconfiguring Affect and Aspirations Through Mobile Communication in Nigeria. *Ethnos*, 83(2): 237–54. doi.org/10.1080/00141844.2015.1120762

Hammar, L. 2010. *Sin, Sex and Stigma: A Pacific Response to HIV and AIDS*. Wantage: Sean Kingston Publishing.

Jolly, M., C. Stewart & C. Brewer, eds 2012. *Engendering Violence in Papua New Guinea*. Canberra: ANU E Press.

Jorgensen, D. 2014. *Gesfaia*: Mobile Phones, Phone Friends, and Anonymous Intimacy in Contemporary Papua New Guinea. Paper presented at CASCA: Canadian Anthropology Society Conference. York University, Toronto, 30 April.

Kelly-Hanku, A., P. Aggleton & P. Shih 2014. 'We Call It a Virus but I Want to Say It's the Devil Inside': Redemption, Moral Reform and Relationships with God among People Living with HIV in Papua New Guinea. *Social Science & Medicine*, 119: 106–13. doi.org/10.1016/j. socscimed.2014.08.020

Lasen, A. 2004. Affective Technologies – Emotions and Mobile Phones. *Receiver*, 11. robertoigarza.files.wordpress.com/2009/07/art-affective-technologiese28093emotionsmobile-phones-lasen-2006.pdf (accessed 14 April 2018).

Pop-Eleches, C., H. Thirumurthy, J.P. Habyarimana, J.G. Ziven, M.P. Goldstein, D. de Walque, L. MacKeen, J. Haberer, S. Kimaiyo, J. Sidle, D. Ngare & D. Bangsberg 2011. Mobile Phone Technologies Improve Adherence to Antiretroviral Treatment in Resource-Poor Settings. *AIDS*, 25(6): 825–34. doi.org/10.1097/QAD.0b013e32834380c1

Silva, S.R. 2012. On Emotion and Memories: The Consumption of Mobile Phones as 'Affective Technology'. *International Review of Social Research*, 2(1): 157–72. doi.org/10.1515/irsr-2012-0011

Tacchi, J., K. Kitner & K. Crawford 2012. Meaningful Mobility: Gender, Development and Mobile Phones. *Feminist Media Studies*, 12(4): 528–37.

Wardlow, H. 2006. *Wayward Women: Sexuality and Agency in a New Guinea Society*. Berkeley: University of California Press.

—— 2008. 'You Have to Understand: Some of Us Are Glad AIDS Has Arrived': Christianity and Condoms among the Huli, Papua New Guinea. In *Making Sense of AIDS: Culture, Sexuality, and Power in Melanesia*. L. Butt & R. Eves, eds. Honolulu: University of Hawai'i Press, pp. 187–205.

—— 2017. The (Extra)ordinary Ethics of Being HIV-Positive in Rural Papua New Guinea. *Journal of the Royal Anthropological Institute*, 23(1): 103–19. doi.org/ 10.1111/1467-9655.12546

3

Toby and 'the Mobile System': Apocalypse and Salvation in Papua New Guinea's Wireless Network

Dan Jorgensen

Introduction

The rapid spread of mobile phones is one of the most striking developments of the last decade in the Global South, where the number of handsets in use outstripped those in the developed world some time ago (Ling and Horst 2011). Pacific nations were relatively late in joining this global surge, but recent growth has been as dramatic there as elsewhere and has attracted the interest of anthropologists and others. There is now an emerging literature examining, for example, the impact of mobile phone use on gender relations, in service delivery and in the various kinds of transactions and links that mobile telephony enables or fosters (see, for example, Gershon and Bell 2013; Marshall and Notley 2014; Lipset, Chapter 1).

Much of what we have come to know about mobile phones in Papua New Guinea has been based on an analysis of patterns of use and their impact on social relations. In this chapter, however, I wish to look at local perceptions of the broader context of mobile phone technology, telecommunications companies and the state. When people in Papua

New Guinea talk generally about the advent of mobile phones, they often speak of 'the mobile system' (*mobail sistem*), a usage that often carries with it an ambivalent sense of larger changes underway. I approach this here through the lens of one man's vision of the dangers that the mobile system offers – his version of the 'big picture' – while juxtaposing this view with equally powerful experiences of mobile phone use in his life. Before doing so, however, I want to sketch a brief history of the recent growth of Papua New Guinea's mobile phone market.

Digicel enters the PNG Market

The growth of mobile telephony in the Pacific follows World Trade Organization–driven regulatory and financial liberalisation that forced state telecommunications monopolies to compete with new corporate players, who often established an early dominance in the market (Duncan 2010; Network Strategies 2013). The most prominent such operator is Digicel, a private telecoms corporation based in Jamaica (with Irish roots).[1] Digicel has made a specialty of expanding into state-run island markets that were too small to attract the attention of larger transnational corporations, and Papua New Guinea offered an outstanding opportunity in this niche (Condon 2008). Describing Papua New Guinea's government-owned Telikom system (the parent of Bemobile) at the time, 'Ofa says:

> As is the case with many state owned entities, the infrastructure had deteriorated over the years as the result of ineffective management, constant political interference, tariffs not being cost-reflective, massive underinvestment and lack of incentives to improve services or roll out network. The infrastructure was outdated, unreliable, and network coverage was low. (2010: 74)

---

1   Digicel Papua New Guinea is a subsidiary of the parent company, Digicel, currently operating in 31 markets in the Caribbean, Central America and Asia-Pacific (Digicel Group Limited n.d.; Horst and Miller 2006; Foster, Chapter 6). In 2014, Bemobile and Vodafone went into partnership to form bmobile–vodafone but it is still too early to assess the impact of this merger on the PNG market.

It's perhaps no surprise that Digicel faced state opposition in 2007 on the eve of its entry into the PNG market – despite the fact that it had won a spectrum-licence auction a year earlier. The minister for telecommunications attempted to revoke Digicel's licence in one of a series of illegal moves to protect the state-owned service provider, Bemobile – ostensibly to allow it time to rehabilitate its infrastructure in order to compete effectively against the newcomer ('Ofa 2010; *PNG Post-Courier* 2007; Radio New Zealand 2007).

The ploy was promptly overturned in court and Digicel lost no time in outcompeting Bemobile through strategies that the company developed earlier in the Caribbean (Condon 2008; Narokobi 2007; Horst and Miller 2006). Digicel went head-to-head with Bemobile for the urban business market and mounted an aggressive advertising campaign (see Horst, Chapter 4), but the key to its plan of attack was to provide mobile phone service to the large majority of relatively poor Papua New Guineans. Cheap handsets (for prices reportedly as low as US$6), low pay-as-you-go rates and special marketing gambits – such as offering free SIM cards or promotions in which customers could trade in their Bemobile units for new Digicel phones at no cost – initially served to secure Digicel's PNG foothold. The number of mobile phone subscribers shot up immediately and some observers credit this expansion with a 0.7 per cent GDP increase the following year (Logan 2012: 1).

Most of Papua New Guinea's population is rural, and Digicel introduced novel features providing a limited number of free 'credit me' or 'out of credit' calls that users could make to other Digicel customers, who could then transfer credits to friends over the phone (see Andersen 2013; Foster, Chapter 6). This innovation was crucial because it inaugurated an informal, flexible and low-cost remittance mechanism that allowed Digicel to leverage the incomes of urban subscribers to subsidise rural friends and relatives at a cost of K0.30 per credit transfer. Rural coverage became profitable, and this helped fund Digicel's tower-building program, which outran Bemobile at a pace that its rival could not – or, at least, did not – match. In what its competitors termed a 'land grab', Digicel built a network of 130 towers in the first year of operation and, by 2015, had a reported 1,088 towers across the country (Condon 2008; Kramer 2016).[2] Papua New Guinea's mobile phone market grew from roughly 300,000 in 2007 to about 3.4 million in 2014 (International Telecommunications Union (ITU) 2015a), and Digicel accounted for nearly all of this growth (Bruett and Firpo 2009). Estimates in recent years have put Digicel's market share at between 90 and 95 per cent.[3]

Digicel's expansion was breathtaking and mobiles went from being the prerogative of 'big shots' to something that even rural schoolchildren have. If not quite a revolution, the advent of accessible phones has been nearly so, and uptake has been enthusiastic. This enthusiasm, however, is often tempered with an uneasiness that is not always readily pinned down. Spontaneous remarks, letters to newspapers and sermons suggest misgivings about the phone's possibilities for fostering sexual immorality, spreading pornography or acting as a mechanism for the harassment of women (Sullivan 2010; Watson 2013; Andersen 2013; Lipset 2013; Jorgensen 2014; Kraemer 2015). To these may be added parents' and teachers' complaints that the phone impairs children's exam performance and the worry that phones could be conduits for sorcery (Bell 2011; see also Pacific Institute of Public Policy (PIPP) 2008:76) or used for criminal purposes. Taken together, these views reveal a surprising ambivalence about a device that most say they cannot do without.

---

2   It is difficult to find reliable accounts of the size and distribution of Bemobile's network in the early days of competition with Digicel, but reference to the environmental assessment report Bemobile submitted to the Asian Development Bank in its bid to secure funding for network expansion (Bemobile Ltd 2011) gives an idea. Bemobile's network consisted of 188 towers in 2011 – four years after Digicel launched its tower-building drive – and it was seeking support for an additional 300 towers over a two-year period. Significantly, this expansion was largely in urban areas and along highways, tacitly conceding more remote rural locations where Digicel had already established a presence.

3   A news report quotes a Digicel PNG executive saying that Digicel's market share was 96 per cent (Papua New Guinea Today 2015b). In 2011, Bemobile estimated that Digicel had over 80 per cent of the market (2011: 3), but the higher figure is consistent with unofficial industry estimates in 2012–13. As one source explained, the likelihood of another carrier wresting Digicel's share away with new tower construction is slim: profitability would depend on persuading enough Digicel subscribers to switch carriers to recoup high construction costs (compare bmobile's coverage map, www.bmobile. com.pg/NetworkCoverage). Horst (Chapter 4) makes a similar argument with regard to Digicel's problems in penetrating the Fijian market.

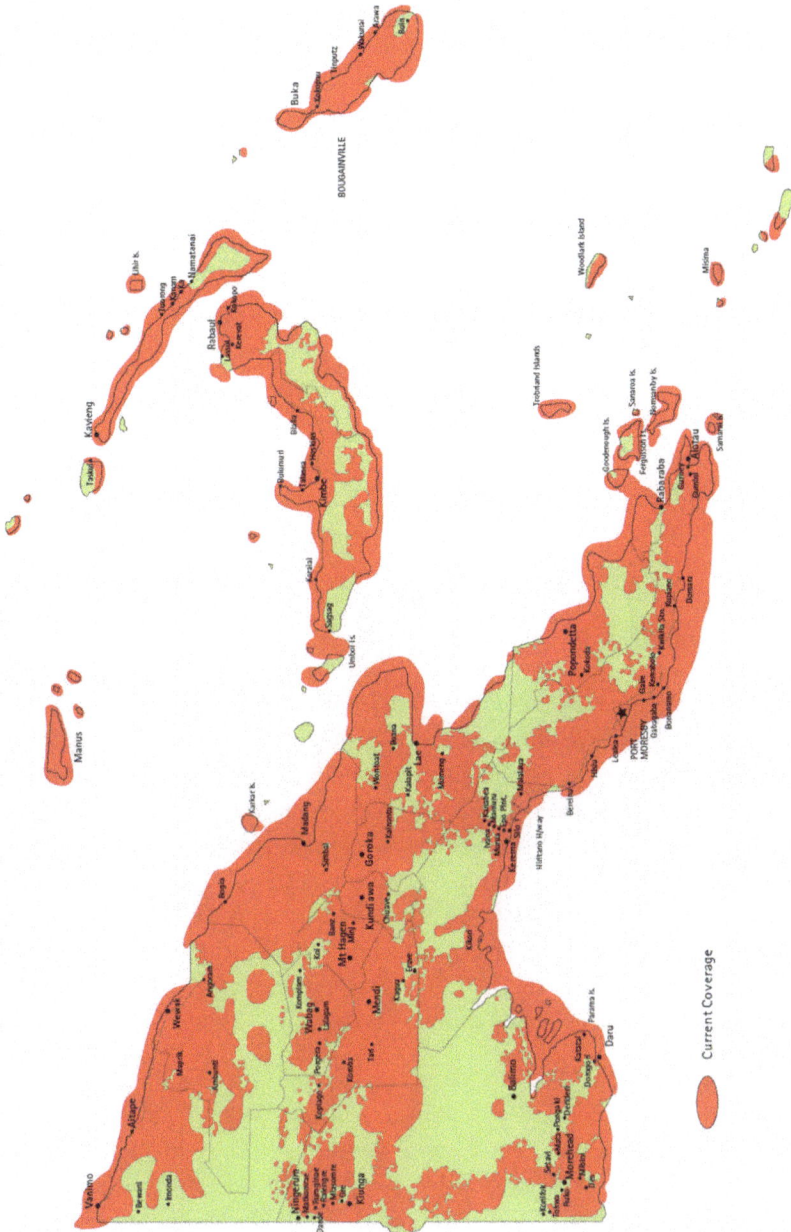

Figure 10. Digicel coverage map, 2015
Source: Digicel Group Limited

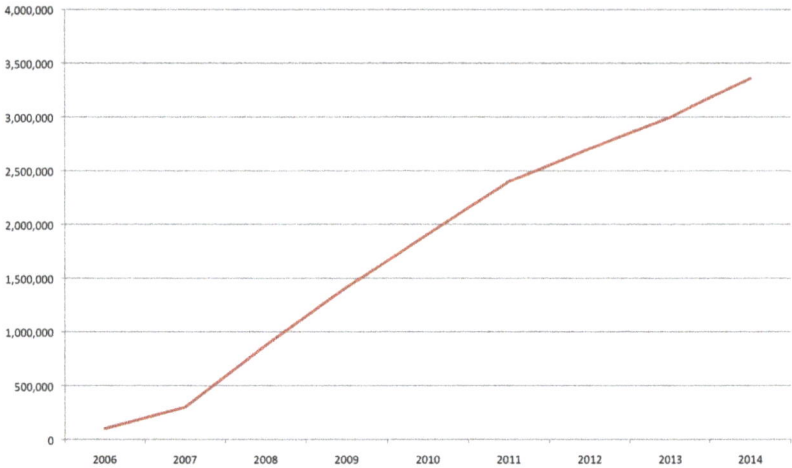

Figure 11. Growth in PNG mobile phone use, 2006–14
Source: ITU (2015)

Toby and the Mobile System

In addition to the specific qualms mentioned above, one sometimes hears that the mobile system is a source of danger in its own right. Some, for example, say that the phone has 'come on a bad road' and is the work of Satan. Here, anticipation contends with apprehension, which I explore through one man's vision of the mobile system's powers and dangers. Drawing upon biblical texts and certain features of state policy, he takes the spread of the mobile system and Digicel's ascendancy as signs of the impending domination of Papua New Guinea by the Americans and Chinese as a prelude to an End Times apocalypse. In his view, this is a story with sinister overtones that is bound to end badly.

I first met Toby in 2014 in a village on the outskirts of Tabubil, a mining town in western Papua New Guinea. I told him that I had come to see what people made of mobiles and whether anything in their lives had changed as a result of their use. Toby, it turned out, was unexpectedly interested in just these sorts of questions. Toby was a migrant whose sister had married into the area and he had followed her there in search of work. He used to work on the docks in Lae, Papua New Guinea's largest port, and there he saw many things, including Digicel's arrival on the scene. Before Digicel, mobile phones were rare, but soon it seemed that nearly everyone had them. At that time, they were fairly simple phones, but Toby noticed that some foreigners had phones that could do much more than just make calls or send texts.

One of these was shown to him by a Chinese crewman, who called Toby over and showed him a video on his phone of his wife and newborn son back in China. He went on to explain that soon one would be able to see people far away using these phones – not in videos, but *live*. Mobiles were very important and would change many things, but most people didn't understand this.

Figure 12. Digicel tower being completed, Telefomin
Source: Stan Tinamnok

In the Highlands near Mt Hagen, Toby explained, a Chinese engineer had been working on a tower that Digicel had just set up. He wanted his family at home to see what kind of work he did, so before he climbed up into the tower he opened his laptop and set it on the ground so that the webcam could send images of him back home. As he got to the top, he looked down and saw four locals converge on his laptop and pick it up, laughing. They were thieves, but the Chinese man began laughing in turn. The thieves asked why he was laughing, and he told them to go ahead and take it – it was rubbish, they were welcome to it. They had seen nothing yet – they could laugh now, but soon enough they would cry for their children. Digicel's towers were extending mobile phone coverage to ever more remote areas of the country and, when the entire country was covered and no place was left out, then they would see. The Americans would be able to see where everyone was: if they wanted to get you, all they had to do was press a button and they could kill you. They could kill you anywhere, and there would be no place to hide. This was the 'mobile system'.

As I later found out, Toby's account was part of a larger vision of events in which Digicel's entry into the PNG market was engineered by US agents who visited Michael Somare – the prime minister at the time – and threatened him in his parliamentary office. There he acquiesced to their demand that the mobile system be extended over Papua New Guinea. The proof of this is in Digicel's rapid expansion and the fact that Digicel forced the government to 'change policy' and assign the number 7 as the initial digit of its SIM cards (ITU 2015b). This signifies that 'you're inside the system now', and refers to Chapter 7 of the Book of Daniel, in which a dream of a 'dragon' and an 'eagle' foretells the roles of China and America in an apocalyptic future.[4] These signs are underscored by the fact that Papua New Guinea's new K100 banknote featured images of parliament on one side and a mobile phone tower on the other, alongside a passenger jet, a truck and a 'submarine'.[5] Toby concludes that:

> All these are things the prophet Amos talked about: he said men will try to get on a ship, or in a car or airplane and go – but there's no road for them to go. Our government said, 'we're the government and we signed this permission saying, United States, you're free to go inside this money now'. Foreigners will think this is okay, but those of us who see this sign know and think: *sori*! The plane's there, the ship's there, but if you board the ship or the plane or get in a car, the Digicel tower will say, 'okay – this is where he is'. And soldiers will get ships, planes, cars and find you – you won't have any way to escape.

---

4   Daniel's dream of Four Beasts is often said to prophesy a future apocalypse and the events of the Book of Revelation. Variants of this interpretation have been staples of certain Christian sects in North America and Papua New Guinea for some time (see End Time Ministries 2014).
5   Toby was technically wrong, since the note had been in circulation since 2005; the 'submarine' is actually an oil tanker (Kamit 2007; Moneypedia 2015).

Figure 13. PNG K100 note[6]
Source: Moneypedia.de

Toby and the Phone in Use

Despite this sinister picture of troubles to come, Toby was remarkably relaxed and upbeat about mobiles and their use. At any given time, he was likely to be using his phone to play video games or listen to FM radio or music that he had downloaded: Toby wearing his earphones was a familiar sight around the village. He was keen to keep up with the latest developments and he often sold a current phone in order to buy a more recent model. He was very proud of the stylish white handset with a touch screen that he was using when we met.

Toby's phone was useful to him in many more personal ways as well. He had come to Tabubil in the hope of finding a way to earn enough money to go back to Lae or to his wife's home village in order to establish a business raising chickens as 'table birds' for Papua New Guinea's frozen food industry. When I met him, he had been living there for over two years and his phone was his lifeline to his wife and to their young daughter back home. Toby also had a 'phone friend' – a young woman elsewhere in the country whose only connection with him was that they had established contact by an anonymous mobile phone call. While they had never met, Toby and his phone friend often shared intimate conversations late at night, when Digicel's rates were reduced.

---

6   The numbers in the figure refer to sectors of Papua New Guinea's national economy, such as cash crops, copra, fisheries, logging, oil, mining and telecommunications.

In all these ways, the mobile phone had become a part of Toby's everyday life, despite his apocalyptic vision of its ultimate significance. More importantly, this vision of the phone also existed side by side with a more powerful and intimate sense of the phone's meaning: it had saved his life.

Despite his best efforts, Toby had not been able to find work and eked out a living through a combination of doing odd jobs for his brother-in-law and operating on the fringes of Tabubil's informal economy. He made occasional forays as a gold buyer purchasing 'dust' from those who panned for gold in the outflow from the Ok Tedi mine site, and sometimes dabbled in what he (without elaboration) termed 'risky business'. He was able to get by, but it was not much of a living, and he was continually frustrated by his inability to save money for his business plans.

After living in the village for a year or so, Toby was upset that his brother-in-law had failed to help him with his plans. Disappointed with life in a village surrounded by strangers, he decided to cross the mountain range behind the village to visit a friend in the neighbouring province who, he hoped, might be more helpful. He set off along the muddy track alone, despite knowing that nobody crossed the mountains by themselves – especially not if they were unfamiliar with the terrain.

By late afternoon, Toby had reached the upper slopes as fog and clouds closed in. The light faded, and it began to rain and, later, to hail: he had been caught in the open without shelter or water. As night fell, he slept on a floor of mud, leaves and tree roots. During the night there was heavy rain with thunder and lightning. For reasons he doesn't understand, Toby removed most of his clothing and curled up trying to sleep. When he woke in the morning, it was cold and he was unable to move. Ants crawled over his body, yet he couldn't stand up or even brush them off: he was going to die.

He slept, he thinks, for much of that day, and recalls a wild dog sniffing around him. At one point he heard voices and felt someone touching him. A couple and their children were returning to their home in the next valley, and had seen his clothes along the track. They searched and found him, and woke him up. They gave him water to drink, a bit of food and a shirt to keep him warm – but they could not get him to walk with them. While they were concerned about their own fate and wanting to move on to reach their valley before nightfall, they said they would try to help him. They asked him who he was and where he had come from and, when he named his brother-in-law, they tried to contact someone for help.

I met Toby about year or so after these events, but when reflecting back on them he stressed that it was his phone-assisted rescue that convinced him that God had a plan for him. Ordinarily, it's difficult to 'find the network' up in that part of the mountains. But, that afternoon, the travelling family was able to get a mobile phone signal and call friends in a nearby village and tell them about Toby's situation. They mentioned his brother-in-law and, from that point on, people patched in a series of calls that reached him at his home in Tabubil. He and Toby's sister went to their village house to see if Toby was there, but found it empty. They followed the track to the foot of the mountain, hoping that Toby would come down or that somebody would bring word. After the better part of a day, however, Toby's brother-in-law rounded up a couple of friends to accompany him up the track. Shouting Toby's name as they went, they eventually found him huddled on the ground. After giving him water, food and a jacket for warmth, they brought him back home.

Connecting Dots

Toby's relation to the mobile system is encompassed in two stories with very different implications. Taking his account of the mobile system and its menace first, it's important to realise that Toby's worries are about Digicel's network but not about the phones themselves, and this suggests that Digicel's claim to be 'the bigger, better network' might provide a clue to his misgivings. In fact, Toby's first story registers worries about Digicel's rise to dominance, raising questions about the nature and sources of its evident power even in the face of government opposition. This isn't unfamiliar territory, nor is it surprising that either Americans or Chinese figure in this context: both countries have been powerful players in Papua New Guinea's resource-dependent economy, with Exxon Mobil's mammoth liquefied natural gas (LNG) project and Chinese acquisitions in the mining industry.[7] As with resource industries, Digicel is foreign, obviously powerful, and nobody is quite sure how far that power extends, or whether it is ultimately sinister or benign (for example, Bobola 2013).

---

7   At the Ramu nickel mine, for example, and at the Frieda River mining prospect. On images of foreigners in Papua New Guinea's globalising economy, see Wood (1995).

The sense that the mobile system is everywhere also suits it as a lightning rod for anxieties concerning globalisation's powerful but often invisible effects and their sources. Such anxieties are staples of the apocalyptic narratives that provide a low background hum in PNG popular culture. Over the years, many of these narratives seized upon digital technology in scares about barcodes, computers, Y2K, microchip implants and, now, mobile phone networks (for example, Bashkow 2000; Jorgensen 2005: 444–45; Bell 2011). End-time narratives cast such technologies as signs and instruments of a coming 'One-World Government' and the tribulations it will bring (for example, Robbins 2001: 542). Quite apart from their role in Christian apocalyptical ideas, what these digital technologies share is the potential (and explicit aim) of connecting the user to unseen others, and this suggests the possibility of surveillance and unwanted intrusions without one's knowledge.[8] Writing of the Asaro Valley of Papua New Guinea's Eastern Highlands, for example, Strong (2015: 6) notes that:

> the building of cellular towers above the valley [is] said to be linked into a one world government and controlled in New York City's new World Trade Center in order to see inside people's homes.

Digicel's claim to be able to 'connect you, anywhere you are, whether near or if you're far' (Hot Croc 2010) takes on an ominous cast in this light.

Here it's important not to dismiss the apocalyptic view without taking note of the issues of unseen surveillance that it references: in a world of digital spying and remote-controlled drone strikes, Toby's concerns have a lot in common with many of ours in the developed world.[9] As if to give substance to Toby's story, the PNG Government has apparently followed the lead of several African countries in legislating the registration of all SIM cards (Telegeography 2015), thereby attempting to tie each card to a user identified by name, photograph, occupation and place of residence.[10] Commonly seen as a way of reining in uncomfortable social media commentary on politicians and policies, this move suggests that the

---

8   In at least one instance, connections with the dead may actively be sought using mobile phones as well (Telban and Vávrová 2014). There are also widespread instances of the desire to be connected to unseen and unknown others in the practice of random dialing, or *gesfaia*, as it is known in Papua New Guinea (Jorgensen 2014).

9   For example, mobile phone assassinations, drone strikes and the use of phones as remote detonators.

10   Details available at PNG eHow (2016).

mobile system is a source of discomfort for the PNG Government as well (Papua New Guinea Today 2015a, *National* 2014; see also, Logan 2012). It also elicited a Facebook response that Toby would recognise:

> No place to hide, time is catching up on us, every individual will be hooked up to the one world system.

All of this helps us make sense of Toby's apocalyptic views, but it doesn't explain why he cheerfully continues to use his phone on a daily basis. In closing, I would like to make two suggestions that might help us put this in perspective. The first is that the mobile system – or any phone network, or the Internet – approximates what Morton (2013) has called a 'hyperobject'. Morton defines hyperobjects as entities that are so massively distributed in space and time as to be effectively non-local – though they have *local footprints*. Among other things, he insists that hyperobjects have a kind of agency that exerts its power as resistance to our intentions – but an agency that is hard to pin down to one element or instance.

Conceived as a hyperobject, the mobile system is both patchy and ubiquitous, yet is invisible and not directly graspable as a whole: 'any local "manifestation" of a hyperobject is not directly the hyperobject' (Morton 2013: 1). Yet, as many of Morton's examples suggest, such objects are difficult to ignore, and elicit attempts to posit a sense of the whole that perforce locates agency *somewhere else*. Morton also suggests that the massive but elusive presence of hyperobjects easily gives way to forebodings of a time to come – a 'signal' of what is yet to happen.

This may seem to offer scant help in understanding how Toby can seem to live so happily in the 'awful shadow of some unseen power' (Morton 2013: 25), but at least part of the answer is that his worries about the mobile system are indeed shadows hinting at something unseen – that is, remote. This reaction will be familiar to alarmists, prophets of doom, as well as patently reasonable campaigners seeking to avoid future disaster – the tendency of even those who acknowledge fears about the future to live in the present. This leads me to my second suggestion, which draws upon Robbins' notion of 'everyday millenarianism' (2001).

For Urapmin Christians expecting the millennium, Robbins argues that the sense of an ending doesn't prevent people from going about the usual business of their daily lives. Instead, he shows that – while being attuned to possible signs of an approaching millennium – Urapmin understand themselves to be 'living in parentheses' between the present and an End that may come at any time. He also proposes that local epistemology normalises a sense of uncertainty about the limits of knowledge, as evidenced in the difference between what one hears and sees. Without being too literal minded, I suggest that Toby shares with Urapmin a healthy awareness of the difference between what one is told and what one experiences, and that it is just this difference that operates in his contradictory attitudes to stories about a possible apocalypse and his crisis on the mountain, where *the phone worked when it wasn't expected to*.

If this is true, then it is tempting to argue that Toby's beliefs are, in Bashkow's terms, 'soft beliefs' that remain speculative and, to an important degree, noncommittal. This captures the disconnect between the picture of a ubiquitous but invisible network and the experience of the phone in hand, where the immediacy of the handheld's usefulness trumps worries about the network's potential or larger meaning. It also affirms the view of another villager concerning the phone's drawbacks and benefits:

> With the mobile you can never escape – but then you can never get lost, either.

For Toby, not getting lost was what counted.

Acknowledgements

This chapter originated in a panel at the 2015 American Anthropological Association meetings organised by Bob Foster and Heather Horst on 'The Moral Economy of Mobile Phones: Pacific Island Perspectives'. I am indebted to them and to Barbara Andersen, Don Gardner, David Lipset, Daniela Kraemer, Margaret Jolly, Jeffrey Mantz, Joel Robbins, Tom Strong, Katja Neves, Paige West and Imke Jorgensen for ideas, suggestions and welcome help. In Papua New Guinea, I profited from conversations with Amanda Watson, the late Nancy Sullivan, Linus Digim'Rina, Michael Akhoy, John Dickinson, Olga Temple, Anwar Soussa, Marina van der Vlies, John Kassmann, Lorna McPherson and Joseph Bariamu. Stan Tinamnok, Kaiku Beksep, Ted Rogers, Georgia Kaipu and Jim Robins were helpful in many practical ways. While Digicel Group has permitted the use of its image and artwork, the views, opinions and research expressed in this chapter are those of the author and do not necessarily reflect the official policy or position of Digicel Group or any of its affiliates and entities. My research was funded by the University of Western Ontario's Academic Development Fund and an International Research Grant. My biggest debt is owed to Toby.

References

Andersen, B. 2013. Tricks, Lies, and Mobile Phones: 'Phone Friend' Stories in Papua New Guinea. *Culture, Theory and Critique*, 54(3): 318–34. doi.org/10.1080/14735784.2013.811886

Bashkow, I. 2000. Confusion, Native Skepticism, and Recurring Questions About the Year 2000: 'Soft' Beliefs and Preparations for the Millennium in the Arapesh Region, Papua New Guinea. *Ethnohistory*, 47(1): 133–69. doi.org/10.1215/00141801-47-1-133

Bell, J. 2011. 'I Love My Digicel'. Paper presented at the meetings of the American Anthropological Association, Montreal.

Bemobile Ltd 2011. Environmental Assessment Report. Proposed Equity Investment and Loan Bemobile Expansion Project (Papua New Guinea and Solomon Islands). Report submitted to the Asian Development Bank, Waigani.

Bobola, E. 2013. Is Digicel Becoming a Bigger, Better Monopoly in PNG? *The Masalai Blog*, 18 December. masalai.wordpress.com/2013/12/18/is-digicel-becoming-a-bigger-better-network-a-worrying-thought/ (accessed 1 December 2015).

Bruett, T. & J. Firpo 2009. *Building a Mobile Money Distribution Network in Papua New Guinea*. Suva: Pacific Financial Inclusion Programme.

Condon, B. 2008. Babble Rouser. *Forbes Magazine*, 11 August. www.forbes.com/free_forbes/2008/0811/072_2.html (accessed 5 July 2017).

Digicel Group Limited n.d. History. www.digicelgroup.com/en/about/history.html (accessed November 2016).

Duncan, R., ed. 2010. *The Political Economy of Economic Reform in the Pacific*. Manila: Asian Development Bank.

End Time Ministries 2014. An In-Depth Look at the Four Great Beasts of Daniel 7. www.endtime.com/blog/depth-look-four-great-beasts-daniel-7 (accessed 8 April 2018).

Gershon, I. & J. Bell, eds 2013. The Newness of New Media. *Culture, Theory and Critique*, Special Issue, 54(3): 259–64. doi.org/10.1080/1 4735784.2013.852732

Horst, H. & D. Miller 2006. *The Cell Phone: An Anthropology of Communication*. Oxford: Berg.

Hot Croc 2010. Digicel PNG 'Network' 2010. Cairns: Hot Croc Advertising. www.youtube.com/watch?v=02AovJNnWyM (accessed 12 May 2016).

International Telecommunications Union (ITU) 2015a. Mobile Cellular Subscriptions by Country. www.itu.int/en/ITU-D/Statistics/ Documents/statistics/2015/Mobile_cellular_2000-2014.xls (accessed 23 May 2016).

—— 2015b. National Numbering Plans: Papua New Guinea. www.itu. int/dms_pub/itu-t/oth/02/02/T02020000A40004PDFE.pdf (accessed 6 July 2017).

Jorgensen, D. 2005. Third Wave Evangelism and the Politics of the Global in Papua New Guinea: Spiritual Warfare and the Recreation of Place in Telefolmin. *Oceania*, 75: 444–61. doi.org/ 10.1002/j.1834-4461.2005.tb02902.x

—— 2014. *Gesfaia*: Mobile Phones, Phone Friends, and Anonymous Intimacy in Contemporary Papua New Guinea. Paper presented at CASCA: Canadian Anthropology Society Conference. York University, Toronto, 30 April.

Kamit, L.W. 2007. Papua New Guinea's New Banknote Series. www.bis. org/review/r071123b.pdf (accessed 17 May 2016).

Kraemer, D. 2015. 'Do You Have a Mobile?' Mobile Phone Practices and the Refashioning of Social Relationships in Port Vila Town. *The Australian Journal of Anthropology (TAJA)*, 28(1): 39–55. doi.org/ 10.1111/taja.12165

Kramer, B. 2016. Which is Better Value, Bemobile or Digicel? www. facebook.com/bryan.kramer.90/posts/1637289709857081:0 (accessed 6 July 2017).

Ling, R. & H. Horst 2011. Mobile Communication in the Global South. *New Media and Society*, 13(3): 363–74. doi.org/10.1177/ 1461444810393899

Lipset, D. 2013. *Mobail*: Moral Ambivalence and the Domestication of Mobile Telephones in Peri-Urban Papua New Guinea. *Culture, Theory and Critique*, 54(3): 335–54. doi.org/10.1080/14735784.2013.8265 01

Logan, S. 2012. Rausim! Digital Politics in Papua New Guinea. SSGM Discussion Paper 2012/9. ssgm.bellschool.anu.edu.au/sites/ default/files/publications/attachments/2015-12/2012_9_0.pdf (accessed 13 May 2016).

Marshall, J.P. & T.a Notley, eds 2014. Communication Technology and Social Life: Transformation, Continuity, Disorder and Difference. *The Australian Journal of Anthropology*, Special Issue, 25(2). doi. org/10.1111/taja.12084

Moneypedia 2015. Papua-Neuguinea: Motive auf den Rückseiten. www.moneypedia.de/index.php/Papua-Neuguinea:_Motive_auf_ den_R%C3%BCckseiten#100_Kina (accessed 13 May 2016).

Morton, T. 2013. *Hyperobjects: Philosophy and Ecology after the End of the World*. Minneapolis: University of Minnesota Press.

Narokobi, E. 2007. Digicel Here to Stay. *The Masalai Blog*, 25 July. masalai.wordpress.com/2007/07/25/digicel-here-to-stay (accessed 13 May 2016).

National 2014. Some Control of Social Media a Must. 14 April. www.thenational.com.pg/some-control-of-social-media-a-must/ (accessed 17 July 2017).

Network Strategies Ltd 2013. Affordability of Mobile Services Hampered by Quasi-monopolies in the Pacific. strategies.nzl.com/industry-comment/affordability-of-mobile-services-hampered-by-quasi-monopolies-in-the-pacific/ (accessed 12 May 2016).

'Ofa, S. 2010. Telecommunication Regulatory Reforms and the Credibility Problem: Case Studies from Papua New Guinea and Tonga. In *The Political Economy of Economic Reform in the Pacific*, R. Duncan, ed. Manila: Asian Development Bank, pp. 63–93.

Pacific Institute of Public Policy (PIPP) 2008. *Social and Economic Impact of Introducing Telecommunications Throughout Vanuatu: Research Findings Report 2008*. Port Vila.

Papua New Guinea Today 2015a. PNG's NICTA Launches Cybercrime Policy. news.pngfacts.com/2015/10/pngs-nicta-launches-cyber-crime-policy.html (accessed 8 December 2015).

—— 2015b. SMS Gate for Service Delivery in PNG. news.pngfacts. com/2015/10/sms-gate-for-service-delivery-in-png.html (accessed 1 November 2015).

PNG eHow 2016. Mobile Phone Sim Card Registration Progresses Well in PNG. 15 June. tech.pngfacts.com/2016/06/mobile-phone-sim-card-registration.html (accessed April 2018).

PNG Post-Courier 2007. Somare's Son Yanks Permit of Competing PNG Telecom. 25 July. www.pireport.org/articles/2007/07/25/somare%C3%A2%C2%80%C2%99s-son-yanks-permit-competing-png-telecom (accessed 17 July 2017).

Radio New Zealand 2007. Revoking License in PNG for Mobile Operator Digicell Unlawful says Commission. 31 July. www.radionz.co.nz/international/pacific-news/171634/revoking-license-in-png-for-mobile-operator-digicell-unlawful-says-commission (accessed 17 July 2017).

Robbins, J. 2001. Secrecy and the Sense of an Ending: Narrative, Time, and Everyday Millenarianism in Papua New Guinea and in Christian Fundamentalism. *Comparative Studies in Society and History*, 43(3): 525–51.

Strong, T. 2015. Policing Witchcraft: Quandaries of Law and Justice in Papua New Guinea Today. Paper presented at the American Anthropological Association meetings, Denver.

Sullivan, N. 2010. *Revised Social Assessment for the PNG Rural Communications Project*. Washington: World Bank.

Telban, B. & D. Vávrová 2014. Ringing the Living and the Dead: Mobile Phones in a Sepik Society. *The Australian Journal of Anthropology*, 25: 223–38. doi.org/10.1111/taja.12090

Telegeography 2015. Ghanaian SIM re-registration to commence by end-2015. www.telegeography.com/products/commsupdate/articles/2015/04/01/ghanaian-sim-re-registration-to-commence-by-end-2015/ (accessed 8 April 2018).

Watson, A. 2013. Mobile Phones and Media Use in Madang Province of Papua New Guinea. *Pacific Journalism Review*, 19(2): 156–75.

Wood, M. 1995. 'White Skins', 'Real People' and 'Chinese' in Some Spatial Transformations of Western Province, PNG. *Oceania*, 66(1): 23–50. doi.org/10.1002/j.1834-4461.1995.tb02529.x

4

Creating Consumer-Citizens: Competition, Tradition and the Moral Order of the Mobile Telecommunications Industry in Fiji[1]

Heather A. Horst

On 1 October 2008, Digicel Fiji launched its new mobile network in Albert Park in the nation's capital Suva. Described by many Fijians as 'spectacular', the company held a free concert that featured Jamaican reggae artist Sean Kingston and New Zealand's reggae band Katchafire. The launch was attended by approximately 60,000 Fijians,[2] a significant

1   This chapter was supported by an Australian Research Council Discovery Project DP140103773, The Moral and Cultural Economy of Mobile Phones in the Pacific. I thank my colleague Robert Foster for constructive comments on the chapter at a pivotal point as well as feedback from seminar participants at the School of Government, Development and International Affairs (SGDIA) Seminar Series at the University of the South Pacific in October 2017, the Digital Ethnography Research Centre at RMIT University in August 2016 and the Department of Media and Communications (MECO) Seminar Series at the University of Sydney in 2016. While Digicel Group has permitted the use of its images and artwork, the views, opinions and research expressed in this chapter are those of the author and do not necessarily reflect the official policy or position of Digicel Group or any of its affiliates and entities. The chapter also does not reflect the views and opinions of Vodafone Fiji or any of its affiliates or entities.
2   The trend of launching with Jamaican and Caribbean music stars has been carried out in Tonga and Vanuatu with Jamaican Shaggy in Tonga and Burmudan Collie Buddz who played centre stage in Vanuatu.

crowd given the population of greater Suva in 2008 (est. 400,000), and Digicel Fiji offered promotions for free phones to the first 100 customers and FJ$75 (US$45) free credit. The company also took out full-page ads in the two main national newspapers, *The Fiji Times* and *Fiji Sun*, a day that is widely recognised in the local newspaper industry as breaking records for advertising income thanks to both Digicel Fiji and the nation's incumbent mobile network provider, Vodafone Fiji. In addition, the press conference for the launch was featured on various nightly news programs.

The launch was followed up with Caribbean-style 'roadshows' where caravans of vehicles with loudspeaker systems moved throughout the country announcing the company's arrival in Fiji and associated promotions to entice new consumers to their mobile network. Throughout the events leading up to and after the launch, Digicel Fiji emphasised its cross-island coverage, especially the expansion of its mobile network coverage in the rural and remote areas of Fiji. The company also proudly highlighted its hard-earned reputation for introducing competition and the subsequent reductions in market prices across their markets in the Caribbean, Central America and South Pacific. Indeed, FijiOne's coverage of Digicel Fiji's launch honed in on Digicel's claim to being 'monopoly breakers' who bring 'open competition' at all costs. As one of the Irish representatives of Digicel Fiji declared during the launch, 'we're not afraid to bring it on' (Fiji Broadcasting Corporation 2008).

Digicel Fiji's emphasis upon breaking up monopolies and reducing prices framed the company as introducing a new moral order to Fiji, one that is widely recognised in global capitalism and resonates with the kinds of discourses of rupture, change and transformation associated with market capitalism (Miller 1997; Miyazaki 2006; Polanyi 1944; Shankar and Cavanaugh 2012; Thompson 1991). This chapter examines the ways in which liberalisation and the promise of competition has shaped the telecommunications landscape in Fiji by analysing the branding strategies leading up to and immediately following the launch of Digicel Fiji in 2008. Through close attention to these campaigns and the discourses of change that surrounded them, I argue that liberalisation transformed Vodafone Fiji and Digicel Fiji from mere mobile telecommunications companies providing products and services into moral actors responsible for articulating their responsibilities towards Fiji and Fijians as consumer-citizens. Focusing upon the different forms of moral order created by companies engaging with state agencies as well as consumers, I begin by outlining the ways in which Digicel framed itself as a monopoly breaker that would disrupt existing moral relationships between the incumbent and consumers by offering better and more widespread coverage and affordable prices. I then turn to the incumbent Vodafone Fiji's efforts to both anticipate and respond to the call for a new moral order (Callon, Méadel and Rabeharisoa 2002; Foster 2007, 2011; Slater 2011). In the final section I examine the ways these market conditions and the moral orders associated with them were depicted to the company's current and future mobile consumers.

Competition and Tradition: Competing Moral Orders

As Thompson has argued, the moral economy is based upon 'a consistent traditional view of social norms and obligations [that are] supported by the wider consensus of the community' (Thompson 1991: 188). This section examines the ways in which Digicel and Vodafone marketed transformations in the availability of mobile services – and their associated moral orders – to the consumers and citizens of Fiji (Rutz 1987). As Banet-Weiser (2012) and others have noted, the emergence of brands and branding cultures moves beyond the shift to commodification to offer immaterial products such as emotion and affect as well personalities and values, factors that have become essential to the inner workings of global capitalism. Through these branding activities, users are transformed into consumers and citizens, or consumer-citizens, who use the products of a particular company and come to identify with the brand that produces or distributes a particular product (Couldry 2004; Livingstone, Lunt and Miller 2007). While efforts to brand do not always result in the effects that media scholars and advertisers imagine, advertising and branding do have the potential to shape aspiration and imagination as well as the material conditions of the intended recipients (Comaroff and Comaroff 2009; Dávila 2008; Horst 2014; Mazzarella 2003). Indeed, as Shankar and Cavanaugh (2012: 359) suggest, 'Message, vital to politics as well as advertising, pairs language with materialised forms of affect and emotion in ways that foster identification'. In what follows, I explore how Digicel and Vodafone attempted to establish themselves in Fiji by creating a moral order through which Fijians could begin to recognise themselves as consumer-citizens.

The Digicel Difference: Competition and Value as Moral Order

In almost every country that Digicel entered in order to create a new market, it worked to quickly become a visible feature of public culture through its ubiquitous billboards, extensive marketing programs and, especially, introduction of prepaid services. The company placed high value on its ability to understand and 'respect' diverse cultures while still being attuned to profit (Creaton 2010). Building upon the marketing strategies of established global telecommunications companies like Vodafone and Orange (Goggin 2006, 2010), in Fiji, Digicel undertook sponsorship of sports teams and music competitions. Not surprisingly, Digicel secured sponsorship of Rugby 7s; however, it began sponsoring the team nearly two years before they were awarded a licence to operate in Fiji. In effect, Digicel invested in the Fiji market long before it officially had a service to offer. In addition, it began negotiating with the National Land Trust Board to establish towers in rural and remote areas, also in advance of receiving its licence, to deliver on its promise to the regulatory authority and its claim to offer the 'bigger, better network'.

Given Digicel's aspirations, its advertising practices at the launch involved a two-pronged strategy. The first was a series of advertisements that stressed the presence and quality of Digicel's rural and remote area coverage. Figure 14, for example, is an advertisement published in *The Fiji Times* in October 2008 that featured what looks like a rural farmer in front of a sugarcane field. He was holding a Digicel network phone and the prominent coverage bars – four full bars – reinforced the text of the advertisement, which draws attention to the network's good or 'best' coverage. The rural scenes in these early advertisements suggested the company prioritised a population that the other network operator (Vodafone) previously ignored in building up a mobile network from the urban centres of Fiji. The second advertisement (Figure 15) was of a smiling Fijian woman – made evident for the viewer by the flower in her hair – on a beach in remote Fiji with her hair blowing in the breeze. She was wearing a tank top and *lavalava* (wrap) while walking and talking on the phone, gesturing the number one with her index finger. This was reinforced in capital bold letters with the phrase 'First Time', which the ad's text explained means that the woman's call went through at the first attempt.

Figure 14. Photo of Digicel newspaper advertisement
Source: *The Fiji Times*, 16 October 2008

Figure 15. Photo of Digicel newspaper advertisement
Source: *The Fiji Times*, 14 October 2008

The second strategy, launched on Fijian television in October 2008, featured Digicel-sponsored Rugby 7s players (Digicel Pacific 2009). The commercial began with a rugby ball branded in red Digicel letters falling towards a green island from the sky and landing in the midst of a group of women and children gathered in a rural village. A rugby player on the outside called for the ball and, once tossed over, it is clear that it's Waisale Serevi and the ball has landed on Gau Island. The ball was then tossed between Serevi, young boys and women before Serevi kicked the ball in the air to Isei Lewaqui in Nadi town, passing over a group of Indo-Fijians who were standing in front of Nadi's famous temple. Lewaqui quickly kicked it to Vereniki Goneva in Tavua town. Goneva tossed it around with Indo-Fijian kids and farmers in the sugarcane fields before kicking it on to Lepani Nabuliwaqa in Lakeba in the Lau Islands. Setefano Cakau caught the ball in Suva city before kicking it to the Yasawa Islands and it eventually landed in the hands of the entire Fiji Rugby 7s team.

Whereas both launch advertisements emphasised widespread coverage, the newspaper ads (figures 14 and 15) stressed quality and value. Value in this case was not only about value for money, as exemplified by the woman whose call goes through without having to pay for a connection fee when the call drops out. It was also about the work of valuing rural populations that historically have not been prioritised in the provision of mobile telecommunication services due to the high costs of supporting remote and rural populations. The images of inclusion in the television commercial also signalled the value placed on Fijian culture. As with rugby participation in Fiji generally (Kanemasu and Molnar 2013), all the players in the videos were *iTaukei*. While present in the video, Indo-Fijians played a role in watching and participating from the sidelines (Besnier 2014).

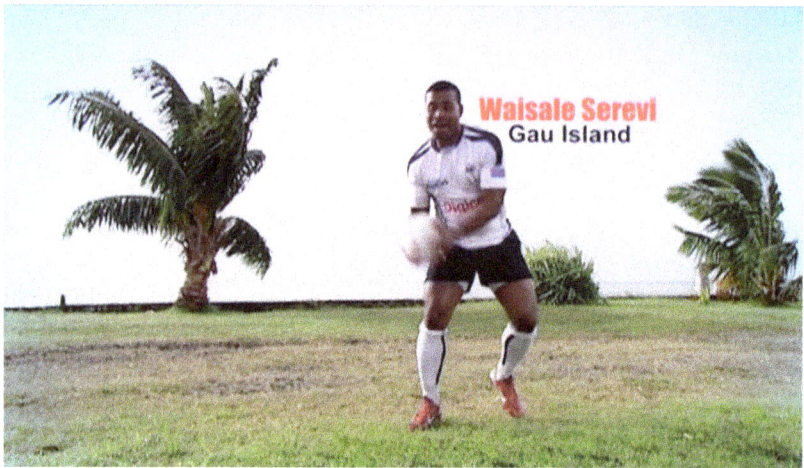

Figure 16. Digicel Fiji commercial
Source: Digicel Group Limited

Figure 17. Digicel Fiji commercial
Source: Digicel Group Limited

The genre of using balls and other paraphernalia from national sports and sports teams is used in advertising globally and the rugby commercial was well received. Fijians recalled that they liked the commercial not only for the celebration of the Rugby 7s team but also because it included a range of men, women and children from different walks of life and emphasised 'ordinary' village and city scenes. They did not, however, read *The Fiji Times* ad of the woman walking on the beach as favourably (see Figure 15). In newspaper editorials, many (self-identified) Christians noted the absence of modesty and believed that the ads should have covered the woman's shoulders and upper arms, as was the case in Digicel's rugby-themed commercial. Others posited that this was an ad directly copied from the Caribbean and was not appropriate for Fiji. Rather than solidarity with other 'island cultures', many Fijians viewed the lack of cultural specificity of these ads as patronising, as if Fijians were another set of 'brown islanders'. While not everyone was offended, the debates around imagery and representation suggest that, at that point in time, value for money and competition were clearly competing with an alternative set of values.[3]

The Value of Vodafone: Time and Tradition

Whereas Digicel Fiji framed its case for a new moral order in terms of the values of competition associated with global capitalism, Vodafone Fiji used the two years between application and launch of Digicel to both counteract Digicel's claim to competition as well as to reframe the negative associations that are often connected with incumbents. This was particularly noticeable in a series of marketing campaigns, the most memorable of which was released just prior to Digicel's launch in 2008: 'Vodafone Fiji Bati Song' (see Mai TV 2008).

This commercial, produced by Art and Soul Fiji, ran for two minutes and 55 seconds and played on major television stations throughout Fiji (see figures 18–21). The video began with a black-and-white image of a man walking barefoot and holding a guitar on what appears to be one of Fiji's pristine white sandy beaches. The only audible sound as the video starts is the washing of the waves until the camera pans to the musician's fingers tuning and strumming the guitar. As the musician plays the guitar, the camera moved to an image of a white flag with the Vodafone symbol blowing in the wind. The video then guides the viewer and listener through a series of timeless Fijian scenes, such as children singing and, of course, Vodafone Fiji rugby players wearing their jerseys. The song's refrain was accompanied by a *bati* (traditional Fijian warrior) leaping through the air

---

3   This continued with responses to the concert at the Digicel launch. A few days after the launch concert there was an editorial in *The Fiji Times* about the negative content of Sean Kingston's lyrics: 'You're way too beautiful girl. That's why it'll never work. You'll have me suicidal, suicidal. When you say it's over.' In particular, various editorials expressed disgust with the lyrics of suicide. Psychologists and others wrote in to recommend that young people speak to pastors and other members of the church for support in similar circumstances.

wielding two drumsticks while a Fijian woman holds a drum. The song concludes with the words 'Spirit of Fiji. We are one. In body, mind and spirit we keep on, we keep strong. Take hold of your destiny from here and beyond. Fiji' while images of Vodafone employees appears between images of rugby players and the song's key vocalist.

Figure 18. Opening sequence with guitar
Source: www.maitv.com.fj (Mai TV 2008)

Figure 19. Vodafone Fiji Flag
Source: www.maitv.com.fj (Mai TV 2008)

Figure 20. Fijian *bati* in the Vodafone commercial
Source: www.maitv.com.fj (Mai TV 2008)

Figure 21. Vodafone Fiji employees singing
Source: www.maitv.com.fj (Mai TV 2008)

Produced by local music artist Daniel Rae Costello, *The Bati Song* was performed by Talei Burns, a talented and much-loved Fijian singer. The song has become a virtual second national anthem and is still routinely played on Fiji One television, although without the Vodafone imagery. As is evident in the description, *The Bati Song* worked to evoke an emotional connection between Vodafone and Fijian citizens through imagery of the Fiji National Rugby Union Team, *batis* and the beauty of nature, as well as spirituality and Fiji's long association with Christianity (Tomlinson 2002, 2009). It also stressed unity and mutual dependence (Brison 2007). The use of black-and-white film for scenes that represent the traditions of Fiji interspersed with contemporary colour images of Vodafone-sponsored teams and employees worked to evoke a sense of timelessness and an understanding of the high value placed on tradition and culture in Fiji, which inspired a sense of pride among most Fijians who watched the video and heard the song.

Consumer-Citizens and Consumer-*Citizens*

From billboards, signs and radio jingles to branded T-shirts, bags and umbrellas, advertisements and associated branding strategies are designed with the aim of creating consumers and developing demand for an evolving suite of mobile-enabled services (for example, Doron and Jeffrey 2013; Goggin 2010; Horst and Miller 2006; Horst 2013; Mazzarella 2003). Advertising of various forms represents one of the most visible features of these relationships between companies and consumers (Arvidsson 2006; Banet-Weiser 2007, 2012; Ciochetto 2011; Lury 2009; Manning 2010; Mazzarella 2003; Shankar 2012, 2015). Indeed, even before its official launch, Digicel secured key billboards throughout the country, erected signs with the Digicel logo welcoming people to villages and began placing Digicel flags on flagpoles throughout the major boulevards of the country. But, as I have demonstrated here, evocative advertisements have as much cachet as the moral order that they seek to tap into and/or create.

Like other countries in the Pacific region (Meese and Mow 2006), liberalisation formally began in Fiji with an open call in 2005 for new mobile telecommunications licences that resulted in the shortlisting of four tenders and the eventual award of a licence to Digicel Limited to operate a GSM network on 26 February 2008 for a 15-year period. Digicel paid a licence fee of US$10.25 million (FJ$15 million) and planned the investment of approximately US$80 million (FJ$120 million) to build the national 2G network. Fiji also became the headquarters of Digicel Pacific. The licence was also accompanied by specific conditions, such as the requirement to focus upon rural areas of the country that were largely undeveloped. This was a requirement by the Telecommunications Authority of Fiji that was designed to guarantee a competitive marketplace for both mobile telecommunications providers. Given their previous experiences in the Caribbean and Central America, Digicel also planned a significant investment in prepaid phone cards that enabled middle- and low-income individuals living in rural and urban areas to control the costs of their calls (see Horst and Miller 2005, 2006).

Vodafone Fiji launched 14 years before Digicel in July 1994 as part of the British-owned Vodafone's third global market in the Asia-Pacific region. It was the first mobile service offered in Fiji. Vodafone's only competition in the telecommunications environment previously was with the national company, Telecom Fiji Limited, which offered landline services to businesses and homes. In 1999, a 49 per cent share of Vodafone Fiji was sold to Vodafone Australia, Vodafone New Zealand and Vodafone Mobile NZ. The remaining 51 per cent stake in Vodafone Fiji was held by Amalgamated Telecom Holdings (ATH), a public company established in March 1998 as a vehicle through which the Fiji Government's investments in the telecommunications sector were consolidated for the purpose of privatisation under its public sector reform program. The ATH ownership made the parent company a minority stakeholder, part of the broader strategy of Vodafone's outreach in the region. Prior to the liberalisation of the market, Vodafone Fiji stood as the sole provider of mobile telecommunications with approximately 41 per cent of the market.[4] Most Vodafone subscribers lived in urban areas and towns on the main island Viti Levu and the second largest island Vanua Levu.

For the two key mobile telecommunications players, then, creating relationships between each of the companies and their consumers, or more precisely *consumer-citizens*, was played out through advertising and branding strategies (Couldry 2004). As a result, the approach used by each of the companies to develop relationships with consumer-citizens

---

4   In July 2014, this ownership arrangement shifted again with the purchase of the remaining 49 per cent stake in Vodafone Fiji Limited by the Fiji National Provident Fund. Purchased for US$87.9 million, the National Provident Fund is the country's main pension scheme that already had a stake in Vodafone. Now the fund has a combined ownership of 79 per cent in the mobile operator. This makes Vodafone Fiji fully locally owned.

varied in subtle ways. Fuelled by liberalisation and the extensive and intensified advertising practices of Digicel Group globally, Digicel's entrée effectively created a commercial public sphere through which the value of consumers, companies and the nation could be negotiated. Digicel framed its moral relationship to consumers in terms of being the impetus to introduce affordable, competitive service across the island. As we see in the early advertisements by Digicel that stressed the rural and remote regions of Fiji's population, the company reframed the Telecom Authority of Fiji's requirement to develop networks in rural and remote regions as an opportunity to bring people living in these regions into the fold as mobile consumers. Like other Pacific and Caribbean markets (Horst and Miller 2006; Meese and Mow 2016), Digicel also focused on prepaid options that appealed to lower-income consumers who were unable to qualify for or maintain postpaid plans.

Vodafone, however, worked to pre-empt and, by extension, undermine Digicel's entrée, effectively chipping away at the efficacy of Digicel's moral commitment to bring down costs and assertions that the company would occupy a role as the 'monopoly breaker' in Fiji. Anticipating the impact of liberalisation on an incumbent monopoly, Vodafone filed an unsuccessful injunction against the government to prevent it from issuing further cellular licences. Alongside the injunction, Vodafone took advantage of the lead time from application to licensing to create a competitor to Digicel. Specifically, in 2007, Vodafone introduced Inkk Mobile Fiji, a subsidiary of Vodafone Fiji that offered low-cost and prepaid phone services aimed at low-income consumers. Inkk Mobile, however, does not possess a separate licence to sell mobile phones and services. While Vodafone views Inkk Mobile as a separate entity, there have also been continual challenges by Fiji's Commerce Commission around confusion of ownership of the company and conflicts of interest, including in 2010 when the Commerce Commission accused Vodafone of engaging in market abuse of power over its relationship with Inkk. From the perspective of Vodafone, which was obviously keen to maintain its market dominance, Inkk Mobile became a prepaid competitor for Digicel.

Moreover, Vodafone presented itself as part of the fabric of Fijian society, leveraging its long-term presence and commitment to Fiji. Vodafone stressed its role as incumbent and market leader by appealing to a sense of belonging, shared history and values during a period of political turbulence and coups (Kelly and Kaplan 2001; Norton 2012).[5] In effect, Digicel's marketing strategies and promotions emphasised the 'consumer' in the *consumer*-citizen equation. Vodafone, by contrast, stressed its relationship with Fijians as consumer-*citizens* through its use of symbols of nation and belonging.

Conclusion

Whereas Digicel's entry into countries such as Jamaica, Samoa and Papua New Guinea has been transformative, resulting in rapid if not exponential rates of adoption (Horst and Miller 2006; Meese and Mow 2016; Watson and Duffield 2016), the uptake of the Digicel network and services in Fiji was less effective. This was largely due to a series of actions taken by Fiji's first mobile company, Vodafone Fiji. Unlike countries where moves toward privatisation disentangled the relationship between the incumbent telecommunications company and the state (Frempong and Atubra 2001; Gao and Rafiq 2009; Gutiérrez and Berg 2000; Howard and Mazaheri 2009), in Fiji the state has maintained a different relationship with the telecommunications companies. Given the ownership structure, it also has important relationships with various regulatory agencies, such as the Telecommunications Authority of Fiji, Fiji International Telecommunications Limited (FINTEL) and the Consumer Council of Fiji, as well as the Fiji National Provident Fund, the local pension scheme that is now majority owner of Vodafone Fiji (see footnote 4). From the perspective of some Fijians, investing in a mobile phone with Vodafone is an investment in one's pension and one's future. Digicel, by contrast, continues to be seen as a 'foreign' entity – albeit one that is now associated as much with the Caribbean as it is with Ireland. Current estimates suggest that Digicel has no more than 20 per cent of the Fijian market, one of the lowest rates of all Digicel's global markets.

For companies, brands function as a source of economic value by cultivating 'loyal customers'. Clearly, however, cultivating consumer-citizens is not just about ads and commercials. We also need to understand what might be pictured as a triangle of relationships among consumers, companies

---

5   Digicel has also engaged in a number of activities to become more local or 'national'. For example, in November 2010, Digicel Fiji changed its logo and applied a shade of blue to the last letters 'cel'. The blue colour represented the background colour of the Fiji flag and coincided with Digicel Fiji's new ad campaign slogan 'Fiji Matters To Us'.

and the state. This might seem an obvious point, but there are very few accounts of mobile telecommunications that bring these three spheres together (although see Doron & Jeffrey 2013). In most anthropological accounts, the state and company represent a backdrop for the 'meaning making' that is evident in the processes of consumption. And, for those in policy, the state and companies set the tenor of the relationship with consumers, who are often relegated to an afterthought. As we see in the case of Fiji, the cultivation of consumer-citizens is as much about the semiotics of emotion and imagery of belonging used in advertising as it is about ownership structures and regulatory bodies. Notions of national belonging are thus intertwined with telecommunications infrastructures.

References

Arvidsson, A. 2006. *Brands: Meaning and Value in Media Culture.* London: Routledge.

Banet-Weiser, S. 2007. *Kids Rule! Nickelodeon and Consumer Citizenship.* Durham, NC: Duke University Press.

—— 2012. *AuthenticTM: The Politics of Ambivalence in a Brand Culture.* New York: New York University Press.

Besnier, N. 2014. Pacific Island Rugby: Histories, Mobilities, Comparisons. *Asia Pacific Journal of Sport and Social Science*, 3(3): 268–76. doi.org/10.1080/21640599.2014.982894

Brison, K. 2007. *Our Wealth is Loving Each Other: Self and Society in Rakiraki Fiji.* Lanham, MD: Lexington Books.

Callon, M., C. Méadel & V. Rabeharisoa 2002. The Economy of Qualities. *Economy and Society*, 31(2): 194–217. doi.org/10.1080/03085140220123126

Ciochetto, L. 2011. Advertising and Value Formation: The Power of Multinational Companies. *Current Sociology*, 59: 173–85. doi.org/10.1177/0011392110391150

Comaroff, J.L. & J. Comaroff 2009. *Ethnicity, Inc.* University of Chicago Press.

Consumer Council of Fiji 2014. Issues Paper: Problems Faced by Consumers in the Mobile Phone Sector. Suva, Fiji.

Couldry, N. 2004. The Productive 'Consumer' and the Dispersed 'Citizen'. *International Journal of Cultural Studies*, 7(1): 21–32. doi.org/10.1177/1367877904040602

Creaton, S. 2010. *A Mobile Fortune: The Life and Times of Denis O'Brien*. London: Aurum Press.

Dávila, A. 2008. *Latino Spin: Public Image and the Whitewashing of Race*. New York: New York University Press.

Digicel Pacific 2009. Digicel Fiji Coverage TV Commercial. *YouTube*, 3 June. www.youtube.com/watch?v=apHTrQUccOY (accessed 14 April 2018).

Doron, A. & R. Jeffrey 2013. *The Great Indian Phone Book: How the Cheap Cell Phone Changes Business, Politics and Daily Life*. Cambridge, MA: Harvard University Press.

Fiji Broadcasting Corporation 2008. *FBC News* (7pm), 1 October.

Foster, R.J. 2007. The Work of the New Economy: Consumers, Brands, and Value Creation. *Cultural Anthropology*, 22(4): 707–31. doi.org/10.1525/can.2007.22.4.707

——— 2011. The Uses of Use Value: Marketing, Value Creation, and the Exigencies of Consumption Work. In *Inside Marketing: Practices, Ideologies Devices*, D. Zwick & J. Cayla, eds. Oxford University Press, pp. 42–57.

Frempong, G.K. & W.H. Atubra 2001. Liberalisation of Telecoms: The Ghanaian Experience. *Telecommunications Policy*, 25(3): 197–210. doi.org/10.1016/S0308-5961(00)00089-6

Gao, P. & A. Rafiq 2009. The Transformation of the Mobile Telecommunications Industry in Pakistan: A Developing Country Perspective. *Telecommunications Policy*, 33(5): 309–23. doi.org/10.1016/j.telpol.2009.03.001

Goggin, G. 2006. *Cell Phone Culture: Mobile Technology in Everyday Life*. London: Routledge.

—— 2010. *Global Mobile Media*. London: Routledge.

Gutiérrez, L.H. & S. Berg 2000. Telecommunications Liberalization and Regulatory Governance: Lessons from Latin America. *Telecommunications Policy*, 24(10): 865–84. doi.org/10.1016/S0308-5961(00)00069-0

Horst, H. 2013. The Infrastructures of Mobile Media: Towards a Future Research Agenda. *Mobile Media and Communication*, 44(1): 147–52. doi.org/10.1177/2050157912464490

—— 2014. From Roots Culture to Sour Fruit: The Aesthetics of Mobile Branding Cultures in Jamaica. *Visual Studies*, 29(2): 191–200. doi.org /10.1080/1472586X.2014.887272

Horst, H.A. & D. Miller 2005. From Kinship to Link-up: Cell Phones and Social Networking in Jamaica. *Current Anthropology*, 46(5): 755–78. doi.org/10.1086/432650

—— 2006. *The Cell Phone: An Anthropology of Communication*. Oxford: Berg.

Howard, P.N. & N. Mazaheri 2009. Telecommunications Reform, Internet Use and Mobile Phone Adoption in the Developing World. *World Development*, 37(7): 1159–69. doi.org/10.1016/j. worlddev.2008.12.005

Kanemasu, Y. & G. Molnar. 2013. Problematizing the Dominant: the Emergence of Alternative Cultural Voices in Fiji Rugby. *Asia Pacific Journal of Sport and Social Science*, 2(1): 14–30. doi.org/10.1080/216 40599.2013.798450

Kelly, J.D. & M. Kaplan 2001. *Represented Communities: Fiji and World Decolonization*. University of Chicago Press.

Livingstone, S., P. Lunt & L. Miller 2007. Citizens, Consumers and the Citizen-Consumer: Articulating the Citizen Interest in Media and Communications Regulation. *Discourse & Communication*, 1(1): pp. 63–89. doi.org/10.1177/1750481307071985

Lury, C. 2009. Brand as Assemblage: Assuming Culture. *Journal of Cultural Economy* 2: 67–82.

Mai TV 2008. Vodafone Fiji Bati Song. *YouTube*, 21 October. www.youtube.com/watch?v=R0W6HyJjjYs (accessed 14 April 2018).

Manning, P. 2010. The Semiotics of Brand. *Annual Review of Anthropology*, 39: 33–49. doi.org/10.1146/annurev.anthro.012809.104939

Mazzarella, W. 2003. 'Very Bombay': Contending with the Global in an Indian Advertising Agency. *Cultural Anthropology*, 18(1): 33–71.

Meese, J. & I.C. Mow 2016. The Regulatory Jewel of the South Pacific: Samoa's Decade of Telecommunications Reform. *Mobile Media & Communication*, 21 February. doi.org/10.1177/2050157916629707

Miller, D. 1997. *Capitalism: An Ethnographic Approach*. Oxford: Berg.

Miyazaki, H. 2006. Economy of Dreams: Hope in Global Capitalism and Its Critiques. *Cultural Anthropology*, 21(2): 147–72. doi.org/ 10.1525/can.2006.21.2.147

Norton, R. 2012. 'A Pre-eminent Right to Political Rule': Indigenous Fijian Power and Multi-ethnic Nation Building. *The Round Table*, 101(6): 521–35. doi.org/10.1080/00358533.2012.749093

Polanyi, K. 1944. *The Great Transformation: The Political and Economic Origins of our Times*. Boston: Beacon.

Rutz, H.J. 1987. Capitalizing on Culture: Moral Ironies in Urban Fiji. *Comparative Studies in Society and History*, 29(3): 533–57. doi.org/ 10.1017/S0010417500014717

Shankar S. 2012. Creating Model Consumers: Producing Ethnicity, Race, and Class in Asian American Advertising. *American Ethnologist*, 39(3): 378–91. doi.org/10.1111/j.1548-1425.2012.01382.x

—— 2015. *Advertising Diversity: Ad Agencies and the Creation of Asian American Advertising*. Durham: Duke University Press.

Shankar, S. & J. Cavanaugh 2012. Language and Materiality in Global Capitalism. *Annual Review of Anthropology*, 41: 355–69. doi.org/ 10.1146/annurev-anthro-092611-145811

Slater, D. 2011. Marketing as Monstrosity: The Impossible Place between Culture and Economy. In *Inside Marketing: Practices, Ideologies Devices*, D. Zwick & J. Cayla, eds. Oxford University Press, pp. 23–41. doi.org/10.1093/acprof:oso/9780199576746.001.0001

Thompson, E.P. 1991. *Customs in Common*. New York: New Press.

Tomlinson, M. 2002. Religious Discourse as Metaculture, *European Journal of Cultural Studies*, 5(1): 25–47. doi.org/10.1177/136494 2002005001154

—— 2009. *In God's Image: The Metaculture of Fijian Christianity*, vol. 5, *The Anthropology of Christianity*. Berkeley: University of California Press.

Watson, A.H.A. & L. Duffield 2016. From Garamut to Mobile Phone: Communication Change in Rural Papua New Guinea. *Mobile Media and Communication*, 4(2): 270–87. doi.org/10.1177/2050157915622658

5

'Working the Mobile': Giving and Spending Phone Credit in Port Vila, Vanuatu

Daniela Kraemer

Introduction

Even if they have only just met, young people in Port Vila, Vanuatu, are quick to ask each other the question: '*Yu gat wan mobael?*' (Do you have a mobile phone?). Inquiring whether a person has a mobile is a euphemism for asking to input a person's phone number into one's network so that it can be called or texted in the future. Asking for a person's phone number, even after just meeting, is common. So much so that, when I asked Alfred, age 22, whether he had a mobile – we were arranging to meet the next day, he exclaimed: '*E kan! Text i fulap!*' (Oh shit! I receive lots of texts!). Alfred fussed that too many people had his number resulting in him receiving, and thus having to decide about responding or not responding to, 'too many texts'.

Most young people in Port Vila own their own mobile, which, like other youth worldwide, they use frequently. Handsets can be seen, and rings and vibrations can be heard, at all hours of the day and night – in town, on the roads, in yards, in the bush and in kava bars. Young people use the phone's flashlight, MP3 player and, of course, its calling and texting functions.

What is most interesting about Port Vila young people's use of the mobile phone technology, however, is that while subscribers in the United Kingdom, United States and Canada use the phone to communicate with known contacts, in Port Vila, the mobile is valued as a tool to build new relationships with unknown contacts. In Vanuatu, people use the mobile phone to phone and text random numbers and start conversations and relationships with strangers. Indeed, by owning a mobile, a person gains access to a social network and all its potential relationships. For young people in Port Vila, mobile telephony is less about communication with family and friends than it is about broadening one's network of social relationships.

Informed by cultural understanding that social success comes through the accumulation of social relationships, young men give and receive mobile phone credit to build and manage their social relationships in an urban context in which they feel marginalised both by kin in town and by their home island–place connections (Kraemer 2013). Indeed, for Alfred and young men like him, the phone has become one of his most important assets and, if it is stolen or if his girlfriend breaks it during a row, as often happens (Kraemer 2015), he will go to great pains to acquire another one. Indeed, as this chapter will show, it is through the mobile phone that many urban young men are able to expand their social life and, with that, their status and renown.

Alfred is one of about 60 young men with whom I spent time during my doctoral fieldwork. Alfred and the other young men were all born in Port Vila and grew up and live in Freswota, one of Port Vila's residential communities. Their ages ranged between 16 and 28 – making them part of the first generation to come of age after Independence in 1980. As they were born and raised in town, their lived experience is remarkably different from that of their parents or grandparents, most of whom migrated to town from smaller rural islands at the time of Independence. As I will discuss later, their engagement with the *kastom* and tradition of their ancestral places as well as with their extended kin is diminished (Kraemer 2013). As such, unlike with previous generations, their social life is defined by the urban community in which they have grown up and the cosmopolitan group of neighbours and peers they live amongst.

***

This chapter explores some of the ways relationships are currently being enacted in urban Vanuatu and the role that is played by mobile phone technology. It looks at how mobile phone consumers use the mobile phone in their quest to experience and create new forms of sociality, as well as the way the technology is being used to enhance the forms of sociality that already exist. For many Port Vila youth, the mobile phone has become the most essential tool used to mediate their relationships. With the new technology comes new opportunities, as well as risks. This chapter thus also explores the ways young people navigate new ideas, values and practices of relating to one another by means of their mobile phones.

New Technology in Vanuatu

Mobile phone technology is relatively new for most ni-Vanuatu. An employee of Vanuatu's first telecommunications company, Telecom Vanuatu Ltd (TVL), placed the first mobile call in 2001. In these early days, TVL employees were encouraged to make random phone calls in order to spark the population's consumption of the new technology. With few mobile subscribers to choose from, the man apparently dialled a random number, which led to a 'nice conversation' with an unknown female. This practice of 'random calling', especially by men looking for women, is a common occurrence in Vanuatu today.

In the early days of mobile telephony, most people could not afford a mobile phone. Statistics from a report published by the Pacific Institute of Public Policy (PIPP), show that, even in 2007 – six years after TVL's network launch – just 11 per cent of the country's population were mobile phone subscribers (Sijapati-Basnett, Brien and Soni 2008: 11). At this time, TVL was the only telecommunication provider in the country and people complained about its high cost and limited network coverage, and pressured the Vanuatu Government to amend the *Telecommunications Act 2006* and open the market to competition (Sijapati-Basnett, Brien and Soni 2008: 11).

Eventually, in December 2008, a licence to operate was issued to Digicel, an Irish-owned company that was already established in the Caribbean and in other Pacific Islands states. Mobile subscriptions skyrocketed as the launch of Digicel broke TVL's monopoly and led to a drop in prices. Sijapati-Basnett emphasises the speed at which the new technology was taken up (pers. comm., 28 Feb 2012). In 2009, the time of the last national census, approximately 76 per cent of all ni-Vanuatu households (rural and town) reported using mobiles (Vanuatu National Statistics Office 2009: 27), and this number has certainly increased since then. As Toto, one of the young men I worked with, once exclaimed: 'Now every place has a mobile phone! Even a person on the islands who has never seen the lights of town, they already have a mobile phone.'

As has been documented in other developing countries, the mobile phone has quickly become the fastest form of communication between people. In Vanuatu, this is the case regardless of a person's social class or educational background because people with little to no regular income can still use the mobile phone's communication services.

The young people I spent time with in Freswota usually only had a few coins in their pockets. Most young women receive small amounts of money from family members for their housework and for looking after the children of family members working in the central business district of town. Young men earn small amounts of money through informal work such as yard work for neighbours, preparing kava for a family member's kava bar, participating in community projects such as building shelters for visiting market vendors or collecting firewood for community ceremonies. Young men also frequently coerce their formally employed girlfriends, sisters and female cousins for money, or resort to theft.

Young men spend less of their money on purchasing mobile phone credit, sometimes called 'talk-time', than young women. Instead, they use a variety of techniques, both improvised and made available by the phone companies, to communicate as much as possible for free. Young women also use these techniques when their purchased talk-time runs out. Similar innovative use of the mobile phone technology for communication has been reported for mobile phone users elsewhere (see Horst and Miller 2006 for Jamaica; Andersen 2011 for Papua New Guinea; Archambault 2012 for southern Mozambique).

In Vanuatu, Digicel allows mobile subscribers who do not have credit to make 'credit requests' of other people. A credit request entails the transfer of phone credit between two phones on the same network. A person requests credit by typing into the mobile *198# then the contact number of the person from whom one wants to 'request credit' and then # and the amount of credit one wants. Usually the requests are for small amounts of phone credit such as 20vt or 50vt (AU$0.24, AU$0.60, respectively). It is both free to send these credit request messages and free to transfer the credit. Youth send credit requests to people whom they suspect have credit in the hope that credit will be shared.

The second technique is the sending of no-cost, pre-written '*plis kol mi* (please call me)' text messages to people with whom they want to speak in the hope that the recipient will use their credit and ring back. Lastly, while it is not a specific feature of the phone's technology, young people who have some credit use the mobile in a creative way – they make a 'missed call' (*mekem mest kol*). In Port Vila, this practice is also known as 'choking the phone' (*jokem fon*). The person rings the number of the person they wish to speak to and then hangs up after one ring. The hope is that when a person sees that their phone has been 'choked', they ring the phone number using their own credit.

Since many of the young people I spent time with in Freswota did not usually have credit, or only had a little credit, these three techniques were used frequently in the hope that someone with credit would send them credit, or the person with whom they wished to speak would have credit or find credit through their own credit requests and then return their call.

When young men like Alfred had credit, they would be particularly savvy about whose please-call-me texts they would reply to and to whom they would give credit. This is because, as the rest of this chapter will explore, their mobile phone etiquette has become one of the most important ways to create and develop relationships and social networks.

Access to Value

One problem faced by many of the young men I worked with is a lack of access to economically valued items. Following Independence, the cost of education became unaffordable for most families and many in this generation had no choice but to leave school – most only obtaining a Year 6 education. Without a high school certificate, many find themselves sidelined from the formal wage economy and unable to earn a sufficient and stable income.

Additionally, these urban young men are disconnected from their home islands. As transportation between islands has become unaffordable for many families in town, parents have not been sending their children to spend time on their home islands as previous generations did (see, for example, Tonkinson 1977; Lindstrom 2011). The consequence is a weakening of urban young people's relationship with the ground of their home island place and with the extended kin of their place (Kraemer 2013). For many of the young men I worked with, being outside this important kinship system means that they do not have access to its support and resources. Moreover, it means that they find themselves outside the customary system of gift giving and the circulation of valued goods. As such, many young urban men find themselves in the problematic and precarious condition of being without the means to establish and maintain enduring supportive social relationships, both in town and in the home islands.

It is not surprising, then, that the young men are not hopeful when they talk about their future. They see themselves as not having the means to move forward in life. Indeed, most of the young men I spent time with, even those in their mid and late 20s, were living with their parents. This was the case even when they had partners and children of their own, as many of them did. Alfred's best friend, 21-year-old Gerome, often lamented this situation. In his words:

> On the island a boy age 21 would have everything. He would have his own garden and yams. He would have pigs. His family would have helped him build a house and soon he would take a wife. But a boy who is 21 years old in town has nothing – no house, no job, no money, no wife, and no chance to have these things in the future. I have nothing and this is why I'm not yet a real man.

Allusions to diminished manhood were common amongst Gerome and his friends. They experience this as an existential problem – seeing themselves as ineffectual men, in a state of permanent youth (Kraemer 2013). For them, it is through participation in economic life that ni-Vanuatu social life is lived and male wholeness is fulfilled.

Anthropological literature has demonstrated the historical significance of gift giving in the formation and management of social relationships in Melanesia. Anthropologists have provided many examples of Melanesians developing relationships and relationship networks through the giving of pigs, mats, yams, kula necklaces and armbands (Malinowski 1966; Munn 1986), and cooked and uncooked foods (Kahn 1986; Munn 1986), amongst other valued items.

The development of relationships through giving has continued with the use of money. In Port Vila, young men and women give friends and family members small amounts of money when they have it. The gift of a small amount of money symbolises the importance of the relationship. One urban tradition demonstrates this explicitly: people in committed relationships give their partner a sizeable portion of their first pay cheque. This act was explained to me in terms of making a 'small contribution' to feed the relationship. Just as with the *kastom* economy, when a person does not give or stops giving, the nature of the relationship is affected. This is especially true in young people's intimate relationships in Port Vila, where not giving some of the money that one has is considered to be an act against the relationship.

Most of the young men I worked with were not formally employed and nearly all had given up on ever finding jobs. While more educated and thus employed urban youth are able to make monetary contributions, or can purchase *kastom* economy to give as gifts to create and maintain relationships, most of the young men I knew did not have this option and experienced this as a significant deficiency in their lives.

Such is the context in which urban young men are exchanging mobile credit and talk time. Informed by cultural ideas and practices in which the exchange of items – be they mats, pigs or money – is key to a person's social development, relationships, status, growth prestige, fame and power, they are exchanging what they do have access to – mobile credit and talk time – in order to develop and manage their social worlds.

As we see, there is more going on in the young men's use of mobile phone credit than using it to communicate or disseminate information. Rather, urban young men are giving and spending mobile phone credit, an item they do have access to, as a way to participate in an exchange system that serves to develop relationships and build and maintain a social network.

Accumulating Mobile Phone Credit and Building up Relationships

To engage in the system, phone credit is needed. Young men like Alfred accumulate phone credit not buy buying it themselves, but by building up a network of people who will respond positively to credit requests they send them – romantically interested young women, employed cousins and siblings, employed neighbours and politicians seeking their support. In a similar vein, they also try to limit how many 'mosquitoes' have access to their number. In Port Vila, 'mosquito' is a term used to describe people who regularly ask others for something but never make a return. These mosquitoes are viewed as consuming other people's sugar or valued items. The young men I worked with were selective about which phone relationships to engage in and, before giving out their number, they consider whether the phone relationship would be taxing, non-reciprocal and unenduring, or worth encouraging and finessing.

Accumulating mobile phone credit is so important to Alfred and young men like him that they pursue it at every opportunity, even stealing it from peers. One way this was done was by discreetly transferring credit off another person's mobile. Alfred was a master of this. He would tell, mostly young women he was flirting with, that he wanted to look at their phones. He would then send a credit request for a small amount of credit from his phone to the woman's phone, discreetly answer the request, and then delete all evidence of this action. Alfred would be long gone before the young woman ever realised that her credit had been syphoned.

Young men go to these extremes to accumulate mobile phone credit because, for them, mobile phone credit has a greater value than the text or talk time it buys – its value is as an exchange item that has become instrumental to social networking.

Once credit is accumulated, young men 'work hard' (*had wok*) sharing or not sharing it with others. Giving mobile phone credit is a way to expand one's network and it is not uncommon for someone to send a stranger credit to instigate a conversation and a relationship. I learned this firsthand not long after I started my fieldwork when I naively gave 15-year-old Tommy, one of my neighbours, my phone number after he asked for it. I wrote in my field notes that night that 'my phone keeps ringing. It's Tommy. First he sent me a 'please call me', then he 'choked' my phone or did a missed call. Then he called 5 times in a row. Then he sent me 100vt credit (AU$1.20) and texted asking whether I had received the credit. It was only when I asked another neighbour, Griffith, to answer Tommy's call that Tommy hung up and didn't call again. Griffith explained that this was an extreme example, that Tommy had just come from the islands and did not know mobile phone etiquette. Griffith said that Tommy's intention with sending me phone credit was to establish a relationship with me. An important point is the finesse of establishing relationships through the giving of phone credit. It is a finesse that, according to Griffith, Tommy, 'fresh from the islands', had not yet learned.

Once a relationship is established, a person can be pressed for money, alcohol, food, clothes and/or additional phone credit to be used to request these items from other contacts. Indeed, Alfred shared his credit with a large network of his peers. They returned his 'generosity' with loyalty and support. Moreover, whenever they happened upon an item of value, such as a bottle of whisky or a bag of marijuana, they would offer Alfred the first taste.

Yet, at least in the initial interaction, the relationships created through giving mobile phone credit, even to strangers, is not based on establishing an obligation on the part of the recipient of the credit to make a return. Rather, as I believe was the case with Tommy, phone credit is given as a tool that works to instigate a relationship that would otherwise not develop, in the hope that this will lead to further gain one day. Horst and Miller (2006) suggest a similar use of phone credit in the context of Jamaica, where Jamaicans give credit to one another not on the expectation of reciprocity but out of a general sense that giving is a further opportunity to 'link-up' – extend and maintain a social connection that has the potential for a positive return in the future. In Port Vila, however, where many urban young men live economically precarious lives, young men are careful not to give phone credit to individuals who they believe will never make a return. This is why Alfred hesitated to give me his number – he had not yet ascertained what kind of phone relationship I would provide. Unlike Tommy, Alfred 'works the mobile' with greater skill.

Modulating Relationships

For Alfred and others, the giving of mobile phone credit has become an important part of developing a social life. Yet, giving and spending credit is not only for building social relationships, it is also a means to modulate relationships as they develop. Indeed, the reason Alfred joked with me about having too many contacts and receiving too many texts is because the more people who have his number, and the more requests he receives, the more he is forced into making explicit those relationships he values and those he does not. As Archambault has written in the context of Mozambique, 'since the entry of mobile phones, one's commitment to a relationship has in fact become easily demonstrable and quantifiable by the money spent' (2012: 39; see also Horst and Miller 2006).

Giving and spending phone credit in a relationship has become a signifier of the value of that relationship to a person. People give to relationships that they want to endure, and do not give to relationships they do not think will be sufficiently supportive or economically or socially profitable. In Port Vila today, it is through phone-mediated giving that people's relationships and enduring commitment are both demonstrated and maintained.

The use of phone credit in this way can be likened to what anthropologists have reported about other gift giving in Melanesia. As Nancy Munn wrote of the Gawan people of Papua New Guinea, certain positively valued acts, such as the giving of food to strangers, or negatively valued acts, such as not giving food, generates the kinds of subjectivities that form the social relationships (Munn 1986: 13). Indeed, among young people in Vila, giving credit creates and activates relationships, while not giving stops relationships from developing.

As I suggest, among unemployed urban young men, phone credit is assigned a value that is more than just its function in allowing for communication. Young people use the giving of phone credit as 'a tangible means to express their relationship with one another' (Taylor and Harper 2003: 159). Phone credit is not given, or is given in calculated amounts, as a means to modulate and maintain a desired degree of intimacy.

A phrase used often by urban young women demonstrates this point. A euphemistic equivalent of the English 'are you still dating him?' is 'does he still ring?' If a boyfriend has stopped ringing, the ignored partner will confront them with the phrase: 'You don't ring anymore! You don't love me'. What this means for Vila young people is that they are evaluating and demonstrating their relationships by the spending or the not spending of phone credit on them.

Throughout Vanuatu, it is socially unacceptable to explicitly deny a request. People generally ignore requests or say that they will think about a request but then never respond. In a similar vein of obfuscation, urban young men use the phrase 'I have zero balance' to reject a request without recrimination. While originally the phrase referred to not having any credit on one's mobile phone, it is now also used as a general term of one's financial state. Saying that one has zero balance allows a person to deny a request and avoid the social recrimination of doing so. Interestingly, the phrase has also become a signifying term of relationships. When a person tells his or her friends that he or she has zero balance, he or she is also acknowledging that his or her ability to make new relationships and manage existing relationships has been momentarily halted.

Masculinity and Renown

Young men also amass credit in order to spend it in acts that assert their masculinity. We see this with the actions of another young man with whom I worked, 25-year-old Toby; although, in Toby's case, his display did not go according to plan. One afternoon Toby recounted to me what had happened earlier that day while he was collecting data for the Vanuatu national census. As he walked around the neighbourhood, a friend asked him to ask the female census worker, Rose, 'whether she had a mobile?'

Toby asked Rose and she wanted to know who was asking. When Toby pointed to his friend, Rose replied that she had a boyfriend. Toby told me that he took offence to her rejection of his friend and decided to retaliate. Toby stole their census supervisor's list to get Rose's phone number. He then traded his T-shirt with another friend for a new SIM card and used it to phone her. Toby said that Rose hung up when she recognised his voice. But, not much later, Toby received a text message from Rose requesting that he send her credit. Toby sent credit and soon his phone rang displaying Rose's number. He answered it, however, it was not Rose's voice it was Rose's brother who reprimanded Toby for disrespecting a woman by stealing her number and calling her.

Toby told me that it got worse. A few minutes later, he received a message asking that he text them a woman's phone number for the brother to call. Toby explained to me that he felt used and so he used swear words in his return text. Having now run out of credit, Toby sent Rose a 'credit request'. Rose texted that she would send credit so they could keep texting but she never did and that was the end of their conversation. Toby told me that this exchange made him cross. He said, 'One day I will find her. She will forget this thing, but I will never forget it and I know her name!'

In this scenario, Toby used his resources to access a new SIM card and phone credit to engage with Rose. It is likely that his plan, as I observed of many young men who feel slighted by young women, was to 'teach her a lesson'. Perhaps he intended to swear at her and insult her as a demonstration of his 'masculine potency'. In any case, what is important here is that his medium of choice was the mobile phone. Had this exchange ended differently, with Toby not appearing foolish, this would have been an example of how young men use amassed credit to assert the masculine power that they experience as diminishing in the urban context.

Toby realised the humour in this story and was not shy about sharing it with me and with his other friends. Indeed, for young men like Toby and Alfred, phone credit is used to instigate fun with their peers. While shared experiences are created through this process, using the phone credit in this way has the added value of creating fame and renown as stories, especially funny ones, spread quickly. Toby and his friends laughed about the incident with Rose and recounted the story to those who hadn't heard it for many days.

The TVL official slogan states: *'storian hemi laef'* (to tell stories is to live). This marketing slogan refers to the history of Vanuatu as an oral society with a strong tradition of storytelling. Young people in Freswota spend much of their time telling stories. Yet, today, the mobile phone is being used not only to communicate stories with others, but to make stories as well. Young people do this by playing practical jokes and tricking people by means of the mobile phone – as occurred between Rose and Toby.

This was also the practice of one of my female friends, Lisa, who was then aged 17. When Lisa realised her boyfriend did not know she had a second mobile phone, she used credit that another friend had sent to her first phone to text her boyfriend under a false name. She teased and flirted with him under this false name until her credit ran out. Throughout she laughed at the fun 'we' were having tricking him like this.

We see here that Toby and Lisa both used their mobile phone credit to entertain people. This is a favourite way for urban young people to spend credit, and young men especially do not keep credit for long but spend it interacting with others. For them, phone credit is to be used to create social relationships, build up a greater relationship network and, through practical jokes using the mobile phone, create shared experiences and stories that potentially bring them renown and fame.

Conclusion

Young people in Port Vila frequently exclaim that their lives have become busy 'working the mobile'. This euphemism points to what I have been discussing – that Port Vila young people, especially unemployed young men, spend time and effort occupied with the accumulation of phone credit, giving or not giving credit, and using credit to build their social networks. They also give credit to modulate their levels of intimacy with community members with whom they have little or no prior social relationship. Giving and spending credit additionally enables them to display their masculine power and garner fame amongst their peers.

For young men like Alfred and Toby, who have learned to accumulate phone credit, finesse 'phone relationships' and give and spend credit, the process often leads to the return of equal or greater value, and enhances their status and renown amongst their peers.

In keeping with previous Melanesian traditions of gift giving, 'working the mobile' has become a form through which young people perform their Melanesian personhood – developing a social life and social success through the act of giving. Indeed, for many of these urban young men, like 'Big Men' before them, giving and sharing, and not giving and not sharing, is the necessary 'work' for a successful social life. Moreover, such innovative use of phone credit demonstrates the resilience and creativity of Port Vila young men who are increasingly marginalised and ignored in their communities.

References

Andersen, B. 2011. Send Credit Please: 'Phone Friends', Financial Mobilities and Gendered Self-Fashioning in Urban Papua New Guinea. Paper presented at the annual meeting of the American Anthropological Association. Montreal, November 16–20.

Archambault, J. 2012. Mobile Phones and the 'Commercialization' of Relationships: Expressions of Masculinity in Southern Mozambique. In *Super Girls, Gangstas, Freeters, and Xenomaniacs: Gender and Modernity in Global Youth Cultures*. K. Brison & S. Dewey, eds. Syracuse University Press.

Horst, H. & D. Miller 2006. *The Cell Phone: An Anthropology of Communication*. Oxford: Berg.

Kahn, M. 1986. *Always Hungry, Never Greedy: Food and the Expression of Gender in a Melanesian Society*. Cambridge University Press.

Kraemer, D. 2013. Planting Roots, Making Place: An Ethnography of Young Men in Port Vila, Vanuatu. PhD Thesis. Department of Social Anthropology, London School of Economics.

——— 2015. 'Do You Have a Mobile?' Mobile Phone Practices and the Refashioning of Social Relationships in Port Vila Town. *The Australian Journal of Anthropology (TAJA)*, 28(1): 39–55. doi.org/10.1111/taja.12165

Lindstrom, L. 2011. Urbane Tannese: Local Perspectives on Settlement Life in Port Vila. *Journale de la Société des Océanistes*, 133: 255–66. doi.org/10.4000/jso.6461

Malinowski. B. 1966 [1922]. *Argonauts of the Western Pacific. An Account of Native Enterprise and Adventure in the Archipelagoes of Melanesian New Guinea*. London: Routledge and Kegan Paul.

Munn, N. 1986. *The Fame of Gawa: A Symbolic Study of Value Transformation in a Massim Papua New Guinea Society*. Cambridge University Press.

Sijapati-Basnett, B., D. Brien & N. Soni 2008. *Social and Economic Impact of Introducing Telecommunications throughout Vanuatu: Research Findings*. Port Vila: Pacific Institute of Public Policy. dfat.gov.au/about-us/publications/Documents/social-economic-impact-of-introducing-telecommunications-throughout-vanuatu-2009.pdf (accessed 14 April 2018).

Taylor, A.S. & R. Harper 2003. The Gift of the Gab?: A Design-Oriented Sociology of Young People's Use of Mobiles. *Computer Supported Cooperative Work*, 12: 267–96. doi.org./10.1023/A:1025091532662

Tonkinson, R. 1977. The Exploitaion of Ambiguity: A New Hebrides Case. In *Exiles and Migrants in Oceania ASAO Monograph No. 5*. M.D. Lieber, ed. Honolulu: University Press of Hawaii.

Vanuatu National Statistics Office 2009. *National Census of Population and Housing*. Port Vila: Ministry of Finance and Economic Management.

6

Top-Up: The Moral Economy of Prepaid Mobile Phone Subscriptions

Robert J. Foster

The term 'moral economy' inevitably evokes Thompson's (1967, 1971) argument that industrial capitalism placed unsettling demands upon workers accustomed to longstanding conventions for setting just prices and following irregular work rhythms. Thompson memorably described how the clock time of methodical wage labour required a new and severe form of thrift – a continuous effort at eliminating wasted seconds and minutes for the sake of discipline and productivity; in short, a Protestant ethic. This historical transition in both time sense and work habits was protracted, conflicted and, probably, never fully accomplished.

I orient this discussion of the use of prepaid mobile phones in Papua New Guinea to Thompson's claims about 19th-century capitalism in places like Manchester textile mills. At the same time, I keep in view Schüll's (2005, 2012) arresting ethnography of 21st-century capitalism as found in places like Las Vegas casinos. New technologies of machine gambling, Schüll argues, encourage habits that are quite different from those we usually associate with Puritan self-discipline. Digital game designers stimulate and enable trancelike consumption without interruption, not only accelerating the pace of play but also extending its duration. Gamers are led, not unwillingly, into a twilight zone where the goal of playing is to continue playing rather than to win. Play, not work, becomes the site for extracting profit.

There is, of course, no radical discontinuity between Thompson's and Schüll's accounts: capitalism remains capitalism. The generation of surplus value is now and always has been a matter of time management – squeezing profit from every available moment. But the socio-technical capacity to extract ever more surplus value from leisure activities such as gambling has brought about the paradoxical situation that I aim to describe here. That is, in Papua New Guinea, it would seem that a recognisably Protestant ethic emerges as a form of self-defence on the part of mobile phone users against the almost incessant corporate incitement to consume.

This chapter ethnographically describes tensions that inhere in company–consumer relations around the mobile phone in Papua New Guinea.[1] It documents how Digicel, the dominant mobile phone service provider, promotes ever new ways for consumers to spend more money on voice calls, text messages and, increasingly, data credits; and it documents how mobile phone users seek ways to avoid paying rent for use of Digicel's self-proclaimed 'bigger and better network'. I argue that prepaid subscriptions can and should be understood as an unstable market device comparable to technologies of machine gambling designed to stimulate game playing and perforce extract greater amounts of revenue from players. This process of extraction paradoxically encourages and undermines temporal discipline; it shapes the practices of mobile phone users, although not without public criticism and creative resistance. By this account, the use of mobile phones signals a larger project of 'managing modernity' (Patterson and Macintyre 2011) in which Papua New Guineans negotiate the terms of their moral and material futures, or so-called development.

***

---

1   This chapter builds upon the conceptual framework of a collaborative research project designed and conducted with Professor Heather Horst (University of Sydney/RMIT University). Our project, funded by the Australian Research Council (DP140103773), is called 'The Moral and Cultural Economy of Mobile Phones in the Pacific'. It involves historical and ethnographic research in Papua New Guinea and Fiji organised in terms of a comparative study of relations between and among companies, state actors and consumers (see Foster and Horst, Introduction).

In Papua New Guinea, the use of mobile phones is underwritten mainly by prepaid subscriptions that enable poor people to purchase mobile phone services on a pay-as-you-go basis (see Donner 2015). Such an arrangement has facilitated the rapid spread and uptake of mobile phones throughout the Global South, where users commonly 'top-up' their air time by purchasing small amounts, often in the form of scratch cards sold by street vendors (Figure 22). Telecommunications companies thus avoid the problem of uncollected bills for services already rendered.

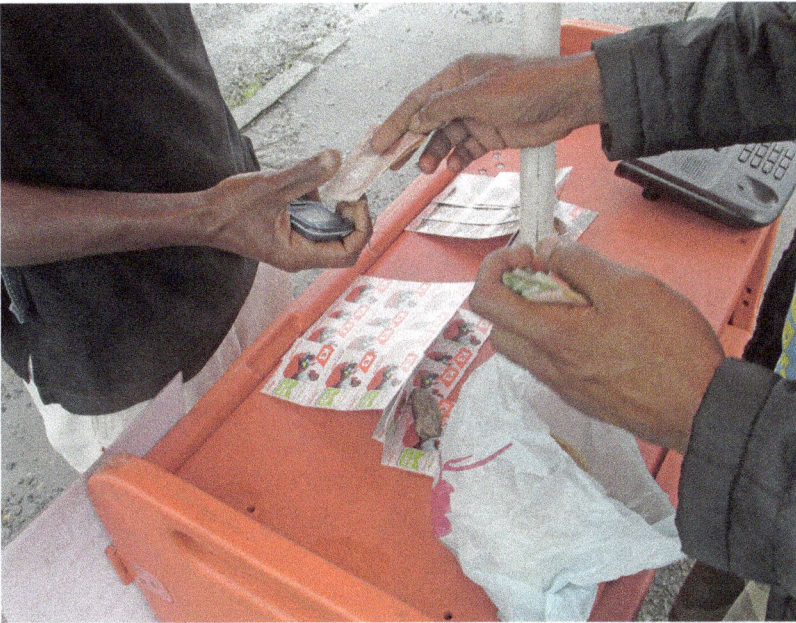

Figure 22. Top-up transaction, Goroka, Papua New Guinea, 2015
Source: Photo by W. Magea

Less acknowledged, however, is how prepaid subscriptions entail a particular moral economy in which mobile users assume fiscal responsibility for managing their phone credits. This responsibility involves dealing not only with the inevitable terms and conditions of use dictated by mobile network operators (MNOs), but also with the frequent enticements issuing from MNOs to increase and accelerate consumption. Put differently, personal discipline is intrinsic to the model of prepayments.

Figure 23. Data bundles in different amounts and durations
Source: Digicel Group Limited

First, discipline is structural. The basic terms and conditions of prepaid subscriptions impose definite constraints. One is of course able to use *only* airtime that one has already paid for, but one also *must* use the airtime one has paid for within a certain period of time or else forfeit it. Digicel PNG airtime credits expire after 30 days, although they might be preserved by topping-up one's balance before the expiration date. Likewise, subscriber identity modules (SIM cards) expire when they have not been used for a certain period of time that might range up to one full year. In Papua New Guinea, Digicel SIM cards expire after three months. It is also worth noting that the sale of data credits for use on laptop computers or mobile phones is also organised in terms of expiration dates. In Papua New Guinea in 2015, one could buy data credits from Digicel that could be used for one hour (10 megabytes (MB) for 99 toea) or for 30 days (1,500 MB for 60 kina) with several options in between (Figure 23).[2] Each option, however, comes with the caveat of 'use it or lose it'. That is, prepayment applies a temporal discipline upon subscribers, one that strongly encourages if not actually enforces consumption. A colleague of mine accordingly confessed that she would stay awake streaming videos on her laptop when she knew that her data credits were about to expire, despite the fact that she had no real interest in the videos and always seemed to regret her decision the following morning.

Second, discipline is subjective, in all senses of the term. Prepaid subscriptions devolve responsibility for managing credits upon the individual using the phone; they encourage the formation of what Donner (2015: 123) calls a 'metered mindset'. One must regularly check the balance of one's account; be aware of the length of calls and whether they are being charged at peak or off-peak rates, or at on- or off-network rates; and be aware of when precisely a data bundle will expire – after which time one's phone credits will be charged at very steep 'out-of-bundle' rates (Digicel currently charges 49 toea per MB). It is up to the individual, not the company, to ensure that he or she has the capacity to make a call or to go online. And that includes, by the way, making sure that the phone's battery is charged, a condition never taken for granted by the vast majority of Papua New Guineans who have uncertain and limited access to electricity (Wardlow, Chapter 2).

Mobile phones, in other words, are devices for promoting a sense of time thrift, an overriding consciousness of the fact that time is money. Lucy, a woman from Bougainville now living in Eastern Highlands Province, told me that Bougainvilleans speak slowly: 'Helllooooo. How are you dooooing?,' she mimicked. She then admitted that, whenever speaking with her family in the village, she is constantly thinking, 'Hurry up! Talk

---

2   100 toea equals 1 kina. 1 kina was equal to US$0.33 in 2016.

quickly! This is costing me!' It is in this regard, I suggest, that mobile phones are effective instruments for teaching an apprehension of time that would please a Puritan – certainly more effective than the discipline of industrial factory labour, which is about as rare these days in Papua New Guinea as it is in the United States. (It is hardly surprising that Digicel was able to create goodwill in many of the Caribbean markets it first entered by offering per-second billing for calls instead of charging by rounding up to the nearest 30 seconds or minute.) Moreover, the sort of personal discipline that mobile phones encourage recalls other technologies – such as self-help programs (Bainton 2010, 2011) and fast-money schemes (Cox 2011, forthcoming) – that Papua New Guineans have embraced in search of financial viability and moral legitimacy.

***

Technologies of prepayment are attracting the attention of scholars working at the intersection of anthropology, geography and science and technology studies (see, for example, Anand 2011, 2014; Baptista 2015; van Heusden 2008; von Schnitzler 2008, 2013). At its most strident, this literature condemns prepayment as a tool of neoliberal governance. Technologies of prepayment – the water and electric meter – depoliticise state–society relations by allowing the state to distance itself from its responsibilities toward citizens. Disconnection from collective services, so the argument goes, becomes self-disconnection – that is, the responsibility of individual 'customers'. These technologies, moreover, are said to inculcate economic rationalities and calculative agency. Von Schnitzler (2008: 902) thus suggests that the prepaid meter, introduced as part of a scheme to privatise water provision in post-apartheid South Africa, is 'central not only to making water calculable, but, more fundamentally, to creating a calculative rationality'. Technologies of prepayment, in other words, entail 'the configuration of economic actors' with specific capacities (von Schnitzler 2008: 902), including the capacity to commensurate social relations with monetary value.

Can any of this be said of prepaid mobile phone subscriptions? I advise proceeding cautiously. There is complexity and nuance in the actual practice of prepaid mobile phone users, and that is what an ethnographic approach promises to highlight. Horst and Miller's (2006) pioneering research on mobile phones in Jamaica, for example, demonstrates that calculability of relationships makes good cultural sense, and certainly is not the result of encroaching marketisation. Quick 'link ups' in Jamaica are normal (Horst and Miller 2005); the fact that an average phone call lasts 19 seconds is not a function of the exigencies of prepayment, but rather the preferred form of keeping multiple relationships open and active.

Do, then, prepaid subscriptions configure economic actors in predictable ways? The everyday use of mobile phones in Papua New Guinea suggests a relationship between consumers and companies that is more messy and less clear-cut than some critiques of neoliberalism would allow. For example, while prepayment does seem to encourage a calculating fiscal responsibility on the part of consumers, the companies providing mobile phone services routinely stimulate subscribers to consume more and more – and in less time. This stimulation, at least in the case of Digicel, derives from the exigencies of continued growth. That is, as the company runs out of new markets to enter in the Global South, it becomes more dependent on a finding a way to get existing users to spend more (Mirani 2013).

In Papua New Guinea and elsewhere, Digicel has endeavoured to put affordable (subsidised) smartphones in the hands of as many people as possible and to offer users a variety of data packages tailored to the budgets of some of the world's poorest consumers. The rapid replacement of basic handsets by smartphones in Papua New Guinea – approximately 700,000 to 800,000 subscribers in early 2016 according to one Digicel official (pers. comm., May 2016) – recalls the analogy with technologies of gambling. Machines designed to both speed up and prolong game playing – for example, by replacing pull handles with push buttons – perforce extract greater amounts of revenue from players. Ditto smartphones. But, as I will show, even users of basic handsets are regularly exhorted to consume. Prepaid subscribers can respond to this stimulation variously – by creatively working around the costs involved in consuming more or by imposing new forms of discipline on themselves *and* on their associates, including friends and family members. What does this dance of stimulus and response look like in everyday practice in Papua New Guinea?

***

I draw here on ongoing fieldwork in Port Moresby (the capital city) and the town of Goroka (in Eastern Highlands Province) that has involved, among other things, formally and informally interviewing a range of approximately 25 mobile phone users (men and women, young and old, rural and urban) and asking approximately two dozen university students to keep detailed diaries of their phone use over a period of 48 hours.

Figure 24. Your family closer: Top-up from overseas
Source: Digicel Group Limited

Research findings overwhelmingly indicate that most people (not just students) have little or no credit on their phone at any given time. Amounts of less than 1 kina are not uncommon. But a low balance or even zero credit does not mean that users are unable to access the Digicel network. In fact, the company enables such access by inviting users to draw on the resources of their social networks by taking advantage of a variety of services. These invitations now encourage subscribers to reach out to friends and family living in Australia and New Zealand, who can go online to top-up the phone accounts of loved ones in Papua New Guinea. Two or three times extra credit is sometimes offered as an incentive and bonus for sending these overseas gifts (Figure 24).

Within Papua New Guinea, users can make a free call to another number to request a return call (Figure 25). This service was imported into Papua New Guinea from the Caribbean, where it was introduced in response to an organically developed version of 'call me' – the missed call. In Jamaica, the introduction of this service in 2004 was welcomed as a clear sign that Digicel understood and respected how Jamaicans communicate with each other.

Figure 25. Call me. Credit me
Source: Digicel Group Limited

In addition, users can send requests for specific amounts of credit from 1 to 99 kina (Figure 25). Credit can be sent directly to another phone for a service charge of 30 toea. Research findings suggest that almost everyone uses this service. Regular credit exchanges of small amounts are a common feature of boyfriend/girlfriend relationships among university students, who also request larger amounts of credit from their parents with similar frequency. Kraemer (Chapter 5; see also Kraemer 2015) has documented the importance of these credit requests and exchanges for the maintenance of social networks among urban youth in Vanuatu, where Digicel also figures as the dominant MNO.

It is also important, however, to note that call me and credit me requests are not always met. (Indeed, the low balances that people keep on their phones can plausibly be interpreted as a strategy for protection against credit requests – a strategy that recalls traditional practices for safeguarding one's resources [betel nut, tobacco, garden produce, shell valuables] – from the claims of others [Foster 1993].) One interviewee thus stated that he does not answer any of the requests that he receives because, if a person has a mobile phone, then that person should be able to afford to operate it. Here, in other words, is evidence of a discourse of self-responsibility taking shape around mobile phones – exactly the sort of discourse critics of neoliberalism might anticipate. At the other extreme, however, Lucy, the woman from Bougainville, claimed that she responds to all requests and even sends credits to friends and family unsolicited. She explained that she was the firstborn in her family and had numerous siblings and other relatives to look out for.

In many instances, people meet requests selectively, deciding when it is appropriate to respond positively and to whom. For example, one woman explained that she responds to her children's call me requests, but not to their credit me requests. She is willing to pay to speak with her children, but not to finance their communication with other people. This response also reflects what some critics of neoliberalism understand as the extension of economic rationality into the domain of the social – that is, the calculation of social relationships in terms of the price of airtime.

Call me and credit me – the company's attempts to extract rent for using the network from someone else if not the actual user – are also met with a kind of creative resistance. Assertions of autonomy – of what one might imagine as the right to communicate – take the form of avoiding service charges or even free riding on the network. For example, one can use free credit requests for specific amounts to send coded messages, many of which have been documented by Olga Temple and her students at the University of Papua New Guinea (Temple et al. 2009). Among the more popular requests are: 99 kina (Goodnight), 43 kina (Love you), 60 kina (Hurry up!) and 24 kina (I have no credit). A request for 24 kina can be sent in polite response to a call me or credit request that one prefers not to accept.

More ambiguous with respect to the question of agency – calculative and otherwise – are the promotions by which the company gives users the means to avoid some but hardly all of the structural limits built into prepaid subscriptions. For example, one much-discussed Digicel promotion rewards users who top-up their balances in the amount of 5 kina or more. A user will receive from Digicel the following text message: 'Congratulations! You have been rewarded with 100min talk time bundle. 100 mins valid for local mobile calls between 11PM–7AM.' These minutes can only be used beginning 11 pm on the day that one tops-up; the bundle will expire at 7 am the next morning.

This particular promotion provokes strange mixtures of discipline and excess that confound any easy answer to the question of what sort of fiscal subjects are constructed by prepaid subscriptions. The notion of time thrift – at least as any Puritan would recognise it – seems perverted when people stay awake until the wee hours of the morning to make phone calls. Indeed, this aspect of the promotion has received public criticism from parents complaining that their children are staying up too late on school nights. I recall one sardonic letter to a newspaper asking whether Digicel thought the author was a flying fox, expecting him to be awake for conversations at 3 am.

On the other hand, time thrift and calculative agency are undeniably present. Users defer making calls during peak-rate periods in order to take advantage of the promotion. One tired young man reported to me that he had received a call from a friend at 1 am. When he anxiously asked the friend if anything was wrong, the friend explained that all was well, but he just did not want to 'waste' the free minutes he received through the promotion. This sort of time thrift risks provoking a negative response for reasons besides disturbing people's sleep. One young woman thus observed that she would not be pleased if her boyfriend called her with free minutes, since it would imply that she was not worth the call otherwise. The currency of phone credits here again measures the value of intimacy.[3]

---

3   This promotion also stimulates the annoying practice of making random calls, sometimes repeatedly to the same number, in search of cross-sex 'phone friends' (Andersen 2013; Jorgensen 2014).

Another popular promotion invites users to purchase a discounted bundle of 60 text messages for K1.20, with the bundle expiring at midnight. Most people with whom I have spoken refer to these text messages as 'free', although economic rationality would insist that each text costs 2 toea (and are advertised by Digicel as such). This particular promotion was one step in a process that illustrates how users are induced to increase consumption. In 2014, a similar promotion offered 50 text messages for 75 toea. (Texts at the time were billed at 25 toea each for the first three, then 6 toea for each text thereafter until midnight; they are now billed at 25 toea per message.) The bundle of 60 text messages thus encourages people to send more texts at a higher price per text. In fact, the promotion encourages the sort of phatic communication in which the message is less important than the act of sending it. Across Papua New Guinea, lots of 'have a top day' and 'good night' messages circulate, especially as the hour of the bundle's expiration draws close. Indeed, some people found 60 texts to be far more than they might need; and others complained that the K1.20 price meant that receiving a gift of 1 kina credit was no longer enough to purchase a text message bundle.[4]

In January 2016, after it was suspended during the Christmas holiday to ensure network stability, the 60 SMS Bundle was relaunched with several new features (Figure 26). The bundle is now valid for 24 hours rather than expiring at midnight. It also auto-renews, allowing the user to carry over unused text messages, unless the user sent a text message to opt out of auto-renewal. It thus becomes the duty of each user to make sure that credit is not unexpectedly deducted from his or her airtime balance. This auto-renew function has also become a standard feature of prepaid data plans, adding one more task to the increasingly complicated management of data on smartphones. While many users complain that their data is disappearing or even being stolen by Digicel (see Foster 2016), the company suggests that users must learn how to manage the settings on their smartphones, turning off the auto-sync function, for example, or remembering that music streaming might use up to 1 MB per minute. That is, Digicel responds to the concerns of its users over how the company measures usage and charges for data with exhortations to assume greater individual responsibility for operating personal devices. Digicel has gone

---

4   In April 2016, I received the following text message several times over the course of a week in Port Moresby, once at 4.57 am: 'The Xpress bundle has arrived! Send Y to 16135 to get 200 SMS + 10MB of Data for K2 valid for 24 hours! Bundle comes with auto-renew. More value with Digicel!'

so far as to offer 'Smartclinics' at its retail outlets in order to educate the public in matters of data usage and social media. The smartphone thus seemingly advances the development of everyday forms of neoliberal governance already entailed in prepaid subscriptions for basic handsets.

60 SMS Bundle

Digicel to Digicel Mobile SMS Bundle: *(Amended and Effective: 19th November 2015)*

60 SMS Bundle now **Auto-Renews** so you carry over your remaining un-used SMS over to your new SMS bundle. Now, the 60 SMS Bundle is valid for 24 hours, that's 24 hours to text all day and night! And you can subscribe multiple times.

Bundle	Price (PGK) GST Inclusive	SMS Count	Validity
60 SMS Bundle	K1.20	60	24 hours

To subscribe to the SMS bundle, send a blank message or the keyword 'SMS' to 1629. SMS Bundle will auto-renew 24 hours from time of purchase. To stop auto-renew, text 'STOP' to 1629.

60 SMS Bundle only applicable for Digi to Digi SMS.

Check your bundle balance at any time by dialing *130#.

Figure 26. 60 SMS Bundle
Source: Digicel Group Limited

***

The main point at stake in examining the practical details of managing prepaid subscriptions is the same one that Mintz (1985) made about sugar in his book *Sweetness and Power*. Demand for sugar is not to be taken for granted, notwithstanding any primate predisposition to consume the stuff. Demand is the outcome of struggles between producers and consumers, struggles that also involve state actors in the form of subsidies, tariffs and trade policies. Much the same can be said for mobile phone services. Demand for mobile phone services is in large measure an outcome of contests in which companies seek ways to undo the fiscal responsibility that consumers learn, ironically, through the pedagogy of prepaid technology.

One logical response to Digicel's promotions and strategies – to its efforts to stimulate more consumption of voice, SMS and data – is greater self-imposed discipline. Let me offer again the example of Lucy, not as typical of Papua New Guineans, but as illustrative of the way in which autonomy and dependence, self-discipline and self-indulgence, merge in the everyday micro-practices of operating a mobile phone. Lucy is a heavy user who can spend up to 100 kina a week in airtime credits. She is a young single woman living far from home and regards frequent communication with her family and friends as nothing less than essential. Lucy, who has a steady income, is aware that she is capable of spending all her savings on airtime, and she has on occasion come close to doing so by topping-up her phone through a mobile banking account with Bank South Pacific – a relatively new service that effectively enables users to top-up any time, any place. (A fair comparison might be made here with gambling machines that provide access to a player's bank account without requiring the player to leave his or her seat in front of the machine.) In order to discipline herself – and Lucy used the English word discipline – Lucy has opened an account with another bank into which she makes weekly deposits. This bank, according to Lucy, does not offer the mobile top-up service. She has thus safeguarded her money from herself.

Lucy is explicit about the calculations that she makes in managing her mobile phone. She says that she does not feel able to start the day unless she is equipped to communicate. So, in the morning, she will top-up her phone for 5 kina. This top-up gives her the aforementioned 100 free promotional minutes to use that night starting at 11 pm. She then purchases a one-day data pass – 60 MB for 3 kina. This data is enough to allow her to go online and communicate with friends and family via the applications WhatsApp and Viber. Lucy only recently discovered that she could send voice messages over the Internet for much less money than making voice calls.[5] Finally, Lucy purchases 60 text messages for K1.20. She will use most if not all of these text messages before they expire at midnight. That leaves 80 toea as a balance in case Lucy needs to make a quick phone call during the day. (On-net calls from one Digicel phone to another are billed at 79 toea per minute during the peak weekday hours of 7 am – 9 pm.) Once she has made these preparations, Lucy feels ready to go out into the world and meet the demands of the day.

***

5   Since I spoke with Lucy, Digicel has changed its pricing policies: 'Effective as of the 15th December 2015, all VOIP (Voice Over Internet Protocal [sic]) calls will be charged at a rate of K1.00 per megabyte.'

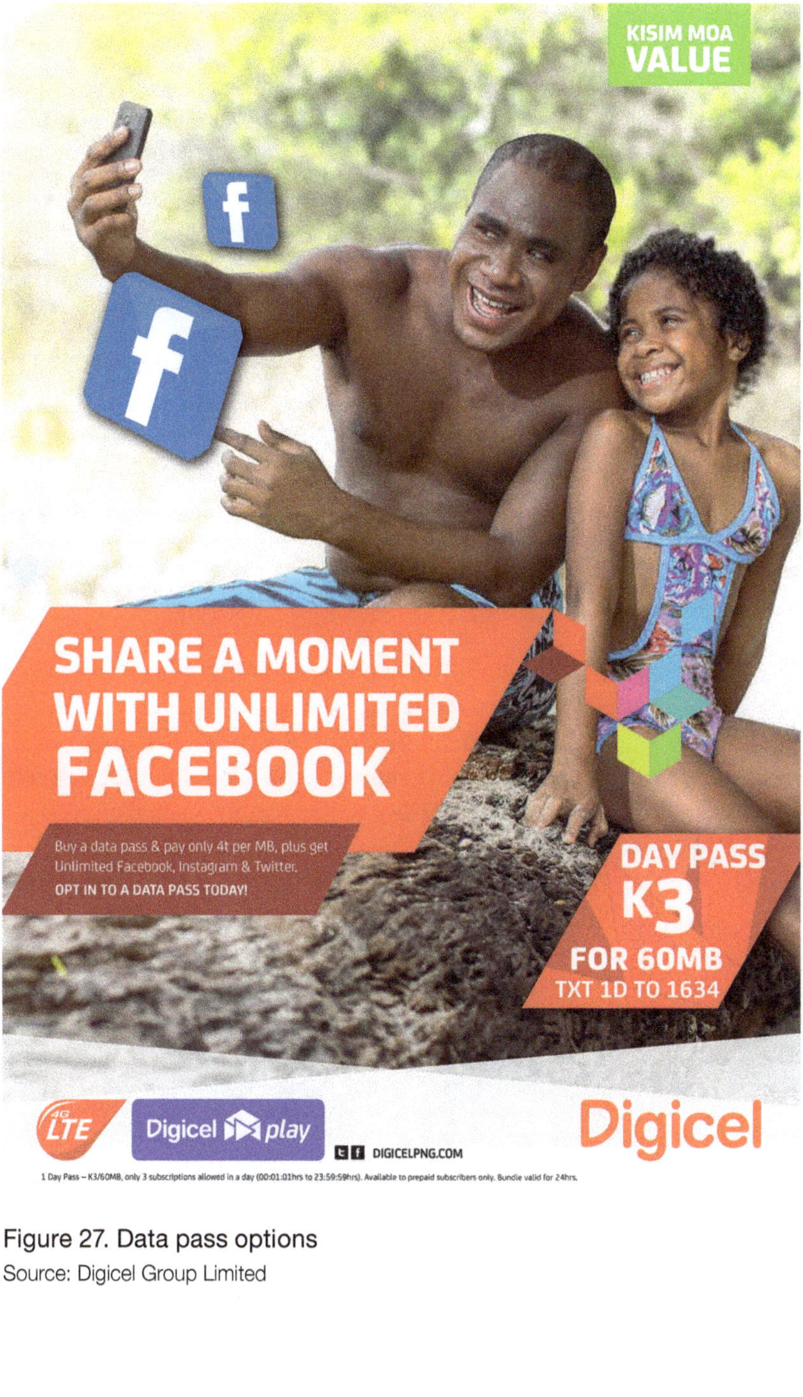

Figure 27. Data pass options

Source: Digicel Group Limited

Thompson talked about the ways in which people became accustomed – more or less – to the new valuations of time that accompanied the Protestant ethic and capitalist work discipline. Prepaid mobile phone subscriptions also subtend new valuations of time, although not in any deterministic way. They are market devices for cultivating responsible consumers – responsible in the sense that these consumers pay for what they consume, and consumers in the sense that they do not simply accumulate airtime and data credits but instead actively use them, top them up and repeat the process as often as possible. Whereas the factory workers that Thompson considered were under pressure to give up all non-purposive activity, the responsible mobile phone consumer is enjoined to chat idly or surf the web. In April 2015, signalling how it imagines the source of future revenues, Digicel accordingly introduced a new data plan: for 4 kina, one can now buy a 24-hour data bundle that supplies not only 60 MB of data, but also unlimited access to Twitter, Instagram and Facebook (Figure 27). In July 2017, Digicel introduced free Facebook basics.

Thompson (1967: 95) imagined a future in which filling one's day with 'leisurely personal and social relations' would be an alternative to and respite from work. Schüll's description of slot machine gambling complicates Thompson's vision; players – retired or, perhaps, on holiday – submit to the soft discipline of digital technologies no less than Manchester weavers conformed to the unrelenting rhythm of power looms. And, as Lucy's morning routine makes clear, enacting personal and social relations is also a form of work, one that requires discipline and that produces surplus value not unlike the labour of a factory hand.

Acknowledgements

Earlier versions of this chapter were presented at seminars at the University of the South Pacific and the University of Papua New Guinea. I thank the audiences at these seminars for their comments and suggestions. I thank the Australian Research Council for funding the research on which this chapter is based and my research partner, Dr Heather Horst, for her steady collegial support. Dr Amanda Watson provided welcome hospitality and generous support for the project. Dr Linus digim'Rina and Dr Verena Thomas helped to coordinate research assistance in Port Moresby and Goroka, respectively. I have benefited from insightful conversations with Amanda Watson, Wendy Bai Magea, Alessandra Mel, Ben Ruli, Alex

Nava and Jason Kariwiga. While Digicel Group has permitted the use of its images and artwork, the views, opinions and research expressed in this chapter are those of the author and do not necessarily reflect the official policy or position of Digicel Group or any of its affiliates and entities.

References

Anand, N. 2011. PRESSURE: The PoliTechnics of Water Supply in Mumbai. *Cultural Anthropology*, 26(4): 542–64.

—— 2014. Consuming Citizenship: Prepaid Meters and the Politics of Technology in Mumbai. Institute for Advanced Study. *Occasional Papers of the School of Social Science*, no. 53. www.sss.ias.edu/files/papers/paper53.pdf (accessed 15 April 2018).

Andersen, B. 2013. Tricks, Lies and Mobile Phones: 'Phone Friend' Stories in Papua New Guinea. *Culture, Theory and Critique* 54(3): 318–34. doi.org/10.1080/14735784.2013.811886

Bainton, N.A. 2010. *The Lihir Dstiny: Cultural Responses to Mining in Melanesia*. Canberra: ANU E Press.

—— 2011. Are You Viable? Personal Avarice, Collective Antagonism and Grassroots Development in Melanesia. In *Managing Modernity in the Western Pacific*. M. Patterson & M. Macintyre, eds. St Lucia: University of Queensland Press, pp. 231–59.

Baptista, I. 2015. 'We Live on Estimates': Everyday Practices of Prepaid Electricity and the Urban Condition in Maputo, Mozambique. *International Journal of Urban and Regional Research*, 39(5): 1004–19. doi.org/10.1111/1468-2427.12314

Cox, J. 2011. Prosperity, Nation and Consumption: Fast Money Schemes in Papua New Guinea. In *Managing Modernity in the Western Pacific*. M. Patterson & M. Macintyre, eds. St Lucia: University of Queensland Press, pp. 172–200.

—— forthcoming. *Fast Money Schemes: Hope and Deception in Papua New Guinea*. Bloomington: Indiana University Press.

Donner, J. 2015. *After Access: Inclusion, Development and a More Mobile Internet*. Cambridge, MA: MIT Press.

Foster, R.J. 1993. Dangerous Circulation and Revelatory Display: Exchange Practices in a New Ireland Society. In *Exchanging Products: Producing Exchange*, Jane Fajans, ed. Sydney: Oceania Monograph Series, no. 43, pp. 15–31.

—— 2016. Digicel Complaints Group: The Moral Economy of Data in Papua New Guinea. Paper presented at the 115th Annual Meeting of the American Anthropological Association, Minneapolis, 17 November.

Horst, H.A. & D. Miller 2005. From Kinship to Link-up: Cell Phones and Social Networking in Jamaica. *Current Anthropology*, 46(5): 755–78. doi.org/10.1086/432650

—— 2006. *The Cell Phone: An Anthropology of Communication*. New York: Berg.

Jorgensen, D. 2014. *Gesfaia*: Mobile Phones, Phone Friends, and Anonymous Intimacy in Contemporary Papua New Guinea. Paper presented at CASCA: Canadian Anthropology Society Conference. York University, Toronto, 30 April.

Kraemer, D. 2015. 'Do You Have a Mobile?' Mobile Phone Practices and the Refashioning of Social Relationships in Port Vila Town. *The Australian Journal of Anthropology (TAJA)*, 28(1): 39–55. doi.org/10.1111/taja.12165

Mintz, S.W. 1985. *Sweetness and Power: The Place of Sugar in Modern History*. New York: Viking.

Mirani, L. 2013. This Company Brought Cell Phone Service to the Remotest Countries on Earth – and Then It Ran Out of Places to Go. *Quartz*. qz.com/154430/this-company-brought-cell-service-to-the-remotest-countries-on-earth-and-then-it-ran-out-of-places-to-go/ (accessed 15 April 2018).

Patterson, M. & M. Macintyre, eds 2011. *Managing Modernity in the Western Pacific*. St Lucia: University of Queensland Press.

Schüll, N.D. 2005. Digital Gambling: The Coincidence of Desire and Design. *Annals of the American Academy of Political and Social Science*, 597(1): 65–81. doi.org/10.1177/0002716204270435

—— 2012. *Addiction by Design: Machine Gambling in Las Vegas.* Princeton University Press.

Temple, O., A. Apakali, D. Bai, D. Dekemba, L. John, G. Matiwat & M. Ginmauli 2009. *PNG SMS Serendipity or sma@upng.ac.pg.* Port Moresby: University of Papua New Guinea Press.

Thompson, E.P. 1967. Time, Work-Discipline, and Industrial Capitalism. *Past and Present*, 38: 56–97.

—— 1971. The Moral Economy of the English Crowd in the Eighteenth Century. *Past and Present*, 50: 71–136.

van Heusden, P. 2008. Discipline and the New 'Logic of Delivery': Prepaid Electricity in South Africa and Beyond. In *Electric Capitalism: Recolonising Africa on the Power Grid.* D.A. MacDonald, ed. Cape Town: HSRC Press, pp. 229–47.

von Schnitzler, A. 2008. Citizenship Prepaid: Water, Calculability, and Techno-Politics in South Africa. *Journal of Southern African Studies*, 34(4): 899–917. doi.org/10.1080/03057070802456821

—— 2013. Traveling Technologies: Infrastructure, Ethical Regimes, and the Materiality of Politics in South Africa. *Cultural Anthropology*, 28(4): 670–93. doi.org/10.1111/cuan.12032

Discussion

Affective Technologies in the Age of Creative Destruction

Jeffrey Mantz

As I read these chapters, I was reminded of Marx's perhaps most important intellectual contribution to understanding the socially transformative power of capitalism: in this modern historical epoch, it is not just things, but time itself (for Marx [1976]) it was our labour time) that becomes commodified. And what others (and objects) do with or to our time tends to bring out the most visceral of human responses. There is a whole subgenre of videos on YouTube featuring people destroying their mobile phones (see Rigg 2013). They drown them, roll over them with construction equipment, douse them in gasoline and set them on fire, pierce them with drills and, of course, microwave them. If affective identities can be expressed through consumption (Miller 1998) in subtle, lingering and disturbing ways, like accessorising your iPhone with colourful absurdities, then it is worth exploring whether the converse might be true. There is something cathartic about blasting the hell out of an object that you feel is tearing apart your being and sensibility. I wonder whether these video bloggers would find common cause and kinship with the masses who smashed all the clocks during the storming of the Bastille (see Benjamin 1968: 261–62). Since the first accountants became the valorised caretakers of modernity, we have been cursed with constantly having to know what time it is. Like many denizens of the new data generation, I have a watch that counts how many steps I take each day, what my average resting heart rate is and how well I have slept. How and when did I become such a numerical fetishist that I care about these totals? These truly are the handcuffs of modernity, and we seem to have fastened them of our own volition.

As Robert Foster notes, 'The generation of surplus value is now and always has been a matter of time management – squeezing profit from every available moment'. That is now firmly entrenched in decision and management science, and the frontiers of efficiency seem to have no bounds. IBM is experimenting with tracking devices for cows in Tasmania to isolate what factors might contribute to optimal dairy production (such as allowing cows to determine the pace at which they are milked). Cowlar, a similar 'Fitbit for cows' technology, is being marketed by a Pakistan-based company.

So the innovation of mobile communications companies using top-up or pay-as-you-go services to capture consumers throughout the Global South is an expected outcome for industries that have to find new markets to survive and thrive. But the disciplining of consumptive behaviour through the use it or lose it phone credit agreements represents an innovative step in this new political economic chapter of market expansion.

We see it elsewhere, of course. Schüll (2012) describes Vegas as a giant machine oriented around extracting profit through its regimenting of time. Time's regimentation is entirely arbitrary and relative. The same processes that enable us to sit in front of a slot machine for 10 hours without the need for a catheter are arguably responsible for our lack of enough patience to watch a viral video clip that is more than six seconds long. As Golub (2004, 2010), Boellstorff (2012, 2015) and many others (e.g. Nardi 2010; Snodgrass et al. 2011) have noted, yes, our online avatars are political economic actors. But, while you can make money selling the things you build in online environments, how long do you have to invest in a massive multiplayer online game to do that?

It is not individual contracts or dollar amounts that matter so much as the habits that are produced and regimented. And new digital technologies have proven astonishingly good at regimenting humans. Elsewhere, I have called the ways that technology 'has subjected its users to forms of social disengagement and disembodiment', a kind of 'zombification' (Mantz 2013: 178). Like a zombie virus, technologies are contagious and depend on an erosion of sensory experience. To put it another way (paraphrasing Lauro and Embry 2008), the social composition of an undead-appearing horde depends on the social decomposition of the humans from which they cull their numbers. Mobile phones are in fact the perfect host for contagion because their very use and existence depends on the user infecting someone else with use. This is what Foster is calling 'temporal discipline', as something that not only encourages but also might 'actually enforce consumption'.

Of course, attempts at the political–economic disciplining of our lives can be met with creative resistance. The sending of free credit requests with coded amounts is an interesting case in point, just like the 'missed call' that pings someone else your location (that they need to call you or come downstairs and meet you, for example). And of course, capital adapts, with 11 pm – 7 am talk time bundles incentivising use at low traffic times, but disciplining sleep (or lack thereof) in the process. As the boundaries between work and leisure are becoming blurred, and what we are calling leisure is turning into work, we have indeed seen an interesting shift from the workers that Thompson discussed, who felt pressure to give up all non-purposive activity. Non-purposive activity is becoming rapidly commodified in this context, and there is an interesting battle over temporality and regimentation of labour taking place. One wonders what Gluckman (1954) might have thought of these pay-as-you-go incentives. Are these the new rituals of rebellion?

We see some of this playing out in Daniela Kraemer's discussion of the reputational economy of mobile phone credits in Vanuatu. Here, mobile phone credits are viewed as an opportunity for young urban men to build relationships outside the *kastom* economy, from which they have been disenfranchised. Social networks are built through the accumulation and redistribution of mobile phone credits. There is no big phone credit *moka* (or other form of ritualised exchange one might find outside Papua New Guinea) at the end of this – it is much more individuated and decentralised – but reputation and prestige are valued in much the same way. As Kraemer notes, drawing on Horst and Miller's (2005) work, this is not about giving with the expectation of reciprocity; it's about the building of relationships and networks. Like the Gawan *kula* ceremonial system described by Munn (1992), credit sharing or the possession of credit, generates the kind of subjectivities that form the social relationships.

One cannot help wondering about issues of gender and class playing out in mobile credit gift economies in ways that they may not play out elsewhere in Vanuatu. The exchange between Toby and Rose is a case in point (Chapter 5). Rose's brother finds the exchange 'disrespectful' of women, while Toby fantasises about retaliating against Rose for stringing him along with the broken promise of a credit transfer. Kraemer asks whether 'teach[ing] her a lesson' is a way of reclaiming his 'masculine potency'. That observation reminds me of the debates about respectability in the island societies of the Caribbean (Besson 1993; Wilson 1973), home to an observably gendered moral and ethical system within which young men expend significant time and energy in building their reputation through any number of acts and investments (certain personal styles, becoming adept in conversation or in skills that draw attention to oneself). These acts and investments become more pronounced as young men are deprived in other economic areas, thus making 'reputation' the most reliable investment to which one should devote resources and time.

Anthropologists working in Caribbean and Oceanic societies have similarly described how vast amounts of time are devoted quite rationally in building non-monetary forms of capital, such as reputation or status. Toby invests here in something that did not represent much financially. But that is precisely the point: this is the form of capital that he can build. Caribbeanist scholars have asked whether these reputational investments are in fact rituals of rebellion. The Caribbean practices that are historically most associated with reputation (Calypso; *mepwi* – a skill that involves one becoming exceptionally adept at insulting others; and sexual conquest, or romantic banditry, if that sounds softer) are all 'skills' that are periodised: they reach a height during Carnival. Here in Vanuatu, we have an arguably 'non-purposive' activity that relies on the estrangement of male youth to facilitate the livelihood of mobile phone credits.

Dan Jorgensen's chapter (Chapter 3) deals with the rise of Digicel's market share dominance in the PNG mobile market, and what social transformations are implied as mobile phone towers begin to blot the landscape at exponential rates. Of course, there are social evils associated with mobile phones. The way they are described here (as associated with moral problems such as sexual impropriety, pornography and harassment) are evocative of the ways that technology has long served as a principle accelerant for immorality. TVs used to be the culprits. Then VCRs eliminated the stigma of having to walk shamefully into adult theatres. The media has changed, but the complexities of social relationships wrought by technological transformation have a long history. Freud (1959) interestingly wrote over 100 years ago about 'modern nervousness', a debilitating ailment of the industrial revolution, best treated through developmental freelisting on his office couch. One might understand this in the context of Papua New Guinea as a crisis of 'respectability' – moral quandaries emerge when the predictability of regimented social relations is thrown asunder. They can even be, as Jorgensen notes drawing on Bell's work, 'conduits for sorcery'. Here, those social evils are mediated. His informant Toby is quite different from Kraemer's Toby. His phone, while it also saved his life, was a vehicle to a different future, where he could get back to a more village-based life in which he and his wife could raise chickens for the PNG food industry. This is why the Chinese crewman story and the apocalyptic warnings to the youths who stole his laptop about the existence of an omnipresent, American-controlled 'mobile system' are so fascinating. Mobile technology for Toby and his ilk is marked by uncertainty. It is a different platform entirely, one that is apocalyptic but simultaneously responsible for saving his life. But he aspires to a future marked less by the oscillation between such radical antipodes, even if he has intimate late-night conversations (presumably with those abundant 11 pm – 7 am talk minutes) with his phone friend. As the villager quoted in the final sentence of his chapter reminds us, 'With the mobile you can never escape – but then you can never get lost, either'.

In rural Sepik villages, mobile phones become devices that reinforce and expand kin relations and social responsibilities, in ways that are accelerated by the phone's inherent capacity to collapse time and space. David Lipset (Chapter 1) notes that initially in Papua New Guinea, attitudes toward mobile phones and the abundant towers were largely positive, reflecting a 'keen desire for connectivity' – remote areas 'explicitly asked to bring in towers'. This desire mirrored the scholarly response: mobile phones increased transparency and accountability and opened lines of communication, much like how some might appeal to the documentary accountability afforded through the installation of body cameras on police officers. The free and open flow of information will save us from ourselves. And there is sociality of course: not just relatives, but also the building of social ties over long distances.

What kind of new 'anonymous modernity' is being formed, which lacks mediation within the dyad of the two doing the communicating, and whose relationship is precarious precisely because it is not mediated (unless Toby's brother gets on Rose's phone, and the brother can remind him of the norms of respectability)? By way of analogy, is this initiation of 'autonomous relationships' responsible for, say, the romantic deserts that younger adults in the cosmopolitan centres of the Global North complain about? Is social media the 'tinder' box (to shamefully reference the social networking and mobile dating app) of enduring social commitments? Unlike urban settings marked already by social disaffection, in rural settings, as Venis reminds us, mobile phones are less about new terrains of sociality and more about reinforcing enduring social bonds and a sense of 'collective responsibility'. The phones of Darapap village reminded me of the single landline phone I found on the college dormitory floor in the 1980s. And, once individual phones were installed in rooms a year later, we all suddenly knew much less about each other. Anyone who happened upon a ringing phone had the responsibility to shepherd the call. Yes, the mobile phone of Darapap can expand the physical space of kinship; and the handset might be private property, but it is not an individualistic device. It may rein in the immoral self, the one that villagers say is engaged in duplicitous behaviour, or infidelity or witchcraft. One does wonder what kinds of calls Venis is shepherding. That phone fosters and reinforces relations beyond the community. As Lipset suggests with his closing sentence, the phone 'may also be associated with the maintenance of moral networks linking rural communities and places with their urban diasporas'.

Heather Horst's (Chapter 4) Fijian mobile phone users are contending with a market environment in which, unlike Papua New Guinea, where Digicel dominates, there is some semblance of competition between two mobile communication giants (Vodafone and Digicel, although a third provider, Vodafone subsidiary Inkk, has a small low-income market share). Horst asks whether the branding strategies deployed in these contexts amount to 'branding cultures', where the experience of buying the product is much more embodied. That's a fair question; we do sleep with our mobile phones after all (though one hesitates to concede too much to a philosophic perspective like that of Georg Simmel [1978], in which our principal form of cultural identification comes to be through the objects that we procure and present to others). My iPhone doesn't define me as a person, but it certainly disembodies a lot of information about me as a person. And all that data is certainly useful to companies that are attempting to profile particular brands and market a viable brand culture. Digicel actively works to develop a brand that can 'understand and "respect" diverse cultures', sponsoring concert venues with hip international stars. Sometimes it backfires when, for example, Sean Kingston or Shaggy says something misogynistic and the press catches wind of it. Their brand of strategic regionalisation seems to have to work harder than Vodafone, which has earned itself a reputation for fostering enduring Fijian futures and traditions, pensions and *bati* songs.

Advertising is indeed often designed to be sentimental and affective in establishing brands, even for unlikely goods and services. American consumers can recall a series of sappy investment banking ads in the 1990s. Thailand now has a series of extended life insurance spots extolling the long-term returns on daily humanitarian investment (for example, an impoverished child whose life is transformed through the modest alms of a good-hearted if monastic benefactor) that are being virally circulated among sobbing global masses. Marketing houses are adept at cultivating sentimentality for some unusual products. But affective marketing doesn't necessarily translate into immediate consumer decisions. Branding is not always about sales; it's about establishing recognition. Coca-Cola certainly does not need to expand its market base in most parts of the world, but they do pay a fee to retain it (and associate it with something that creates an affective attachment, like a generalised reciprocity of soft drinks and smiles). I did wonder whether the highly nationalist approaches to branding was part of a strategy of skirting ethnic differences among Fijians. So Digicel is 'trying', while Vodafone does not have to, because they deal with common national investments like rugby (just as cricket or baseball or soccer become a unifying national force in other parts of the world; see, for example, James 1983). Thus, Horst's call for us to look more at mobile phones as mediated by spheres of consumers, companies and the state is essential to understanding how companies are branded.

Holly Wardlow's chapter (Chapter 2) explores mobile phones as an affective technology, which 'not only mediates the expression and experience of emotion, but also opens up new forms of emotional intimacy', against the backdrop of public health campaigns that view the mobile phone as merely a means of disciplining medication practices. Women with HIV in Tari, Papua New Guinea, use their mobile phones to recruit '"phone friends" who may provide emotional comfort, mental diversion, romantic escapism and even material support'. Similar to what Jorgensen described in the case of Toby, these phone friends are essentially intimate strangers with whom one establishes a friendship through the phoning (and the recipient accepting) a random and unknown call. In exploring mobile phone use in this way, Wardlow resists the typical analytical tendency to treat the mobile phone as a biopolitical/biomedical instrument and instead repositions it as a tool for embodied and agentive practice among an otherwise stigmatised group of women. Its creative use against methods of regimentation and discipline (both at the state level and in other social and institutional organisational contexts) can be extrapolated to other research contexts. Rather than uniquely PNG phenomena, the mobile activities of these women would seem to be analogous to other practices where technology has been used to seek intimate social relationships anonymously. Can this practice be generalised to other practices where unencumbered social relationships are pursued precisely because of the constraints of kin or familial obligations? Social relationships often weigh heavily, particularly for marginalised individuals, and tend not to provide any kind of intimacy. I found it interesting (especially in the cases of Angela and Lucy) how God mediated discussions between phone friends. These conversations seem to function as a sort of testimonial or confession for those whose social order has failed them. If protracted conversations with mobile phone strangers seem on the surface to be a complete waste of time, maybe that is precisely the point: the phone friend provides the antidote to the temporal discipline inherent in other forms of social communication.

These chapters collectively help us to understand the temporal battles that are being waged throughout Oceanic societies as they engage new digital communications technologies. So much of what is discussed in this volume focuses on the how these technologies are embodied in human behaviour. I began my comments with an account of the visceral experiences recounted in the videos of people who delighted in smashing their mobile phones. I did so because we also must remember that human behaviour can become embodied in the physical technologies themselves. Our moral engagement with technologies and the actions that result from such engagement are unpredictable, challenging our ability to control them, lest they come to control us. Nobody is happy when they drop their mobile phone in the toilet. But it sure feels good to throw one on the ground.

References

Benjamin, W. 1968. *Illuminations: Essays and Reflections*. New York: Schocken Books.

Besson, J. 1993. Reputation and Respectability Reconsidered: A New Perspective on Afro-Caribbean Peasant Women. In *Women and Change in the Caribbean*, J.H. Momsen, ed. Indiana University Press, pp. 15–37.

Boellstorff, T. 2012. *Ethnography and Virtual Worlds: A Handbook of Method*. Princeton University Press.

—— 2015. *Coming of Age in Second Life: An Anthropologist Explores the Virtually Human*. Princeton University Press.

Freud, S. 1959 [1908]. Civilized Sexual Morality and Modern Nervous Illness. In *The Standard Edition of the Complete Psychological Works of Sigmund Freud*, vol. 9, J. Strachey, ed. London: Hogarth Press.

Gluckman, M. 1954. *Rituals of Rebellion in South-East Africa*. Manchester University Press.

Golub, A. 2004. Copyright and Taboo. *Anthropological Quarterly* 77(3): 521–30.

—— 2010. Being in the World (of Warcraft): Raiding, Realism, and Knowledge Production in a Massively Multiplayer Online Game. *Anthropological Quarterly* 83(1): 17–45. doi.org/10.1353/anq.0.0110

Horst, H.A. & D. Miller 2005. From Kinship to Link-up: Cell Phones and Social Networking in Jamaica. *Current Anthropology* 46(5): 755–78. doi.org/10.1086/432650

James, C.L.R. 1983. *Beyond a Boundary*. Durham, NC: Duke University Press.

Lauro, S.J. & K. Embry 2008. A Zombie Manifesto: The Nonhuman Condition in the Era of Advanced Capitalism. *boundary 2*, 35(1): 85–108. doi.org/10.1215/01903659-2007-027

Mantz, J.W. 2013. On the Frontlines of the Zombie War in the Congo: Digital Technology, the Trade in Conflicted Minerals, and Zombification. In *Monstrous Cultures in the 21st Century: A Reader*, M. Levina & B.-M. Bui, eds. New York: Continuum Press, pp. 177–92.

Marx, K. 1976 [1867]. *Capital: A Critique of Political Economy, Volume I*. Ben Fowkes, trans. New York: Penguin Books.

Miller, D. 1998. *A Theory of Shopping*. Ithaca, NY: Cornell University Press.

Munn, N.D. 1992. *The Fame of Gawa: A Symbolic Study of Value Transformation in a Massim (Papua New Guinea) Society*. Durham, NC: Duke University Press.

Nardi, B. 2010. *My Life as a Night Elf Priest: An Anthropological Account of World of Warcraft*. Ann Arbor, MI: University of Michigan Press.

Rigg, S. 2013. Comments to Performance and New Media Panel for Workshop, Unseen Connections in the Ecologies of Cell Phones, Smithsonian Institution, Washington, DC, 27 February–1 March.

Schüll, N.D. 2012. *Addiction by Design: Machine Gambling in Las Vegas*. Princeton University Press.

Simmel, G. 1978. *The Philosophy of Money*. Tom Bottomore and David Frisby, trans. London: Routledge & Kegan Paul.

Snodgrass, J.G., M.G. Lacy, H.J. Francois Dengah & J. Fagan 2011. Enhancing One Life Rather than Living Two: Playing MMOs with Offline Friends. *Computers in Human Behavior*, 27(3): 1211–22. doi. org/10.1016/j.chb.2011.01.001

Wilson, P.J. 1973. *Crab Antics: The Social Anthropology of English-speaking Negro Societies of the Caribbean*. New Haven, CT: Yale University Press.

Transforming Place, Time and Person?: Mobile Telephones and Changing Moral Economies in the Western Pacific

Margaret Jolly

The chapters in this volume traverse the countries of Papua New Guinea, Fiji and Vanuatu and relate questions about the political economy and regulatory regimes of companies like Digicel and Vodafone in these three states to the embodied experiences of consumers: young and old, rural and urban, female and male. They are consummately connected (excuse the bad pun). I will focus on three dimensions of this connection by looking at the way in which these chapters reveal how mobile phones have transformed the experience of place, time and person, and how they are situated in changing moral economies.

Place

All chapters attest to the rapid and extensive rollout of mobile telephones across the region. For instance, in Papua New Guinea, the largest country in the region, Digicel expanded from c. 300,000 customers in 2007 to c. 3,400,000 in 2014, and has established about 1,800 signal towers across the country and holds c. 90–95 per cent of the market (Dan Jorgensen, Chapter 3). Holly Wardlow (Chapter 2) observes that, between her departure in 2006 and her return in 2010, villages in the Tari Basin (PNG Highlands) moved from having no mobile phones to having extensive coverage and ownership. In this period, Digicel was expanding at an astonishing rate, compared to its state rival Bemobile, which was hampered by ageing infrastructure, state inertia and messy politics. Vaulting fast in this race with their rival, Digicel used many techniques honed by its experience in the Caribbean that were designed to entice not just the urban rich but the rural poor with cheap phones and systems designed to accommodate the low credits of most users (like the free 'call me' and 'credit me' systems).

Mobile phone coverage in many parts of Papua New Guinea (as in much of rural Vanuatu and Fiji) has preceded the delivery of running water, sewage or electricity. Thus, in many rural regions, mobile phone batteries can only be recharged at available public outlets (e.g. in airports, stores or government departments), at improvised charging stations at markets or by using solar power. As the map that accompanies Jorgensen's chapter shows, and David Lipset (Chapter 1) observes for Murik Lakes in the Sepik, coverage is not total – there are still some villages both on the mainland and offshore islands of Papua New Guinea that are not regularly on the grid. But, compared to Bemobile's coverage, which is defined by major highways and trunk roads, Digicel has penetrated into much of the rural and remote highlands and islands of the nation. The company's rapacious desires to expand their coverage, market dominance and consumer allure has been matched by a strong desire on the part of many of Papua New Guinea's citizens and consumers to be connected. This passion for connection is equally felt in Vanuatu and in Fiji (see Horst , Chapter 4; Kraemer, Chapter 5; Taylor 2015).

But, we might ask for all three countries, how far has this transformed or even revolutionised notions of *ples*: ideas of national citizenship in relation to ethnic and regional differences and boundaries of belonging?

In her chapter on Fiji, Heather Horst evokes not only a contest between the rival companies of Vodafone and Digicel but also a contest between competing boundaries of belonging, from the local to the national to the global. These are not so much *nesting* scales but *contesting* scales of place. She focuses on how the longer established Vodafone, which still commands most of the Fiji market (c. 80 per cent), has deployed a series of branding strategies that evoke a sense of belonging to or being part of Fiji: to the Fiji National Rugby Team, *bati* or Fijian warriors, the natural beauty of the islands, local popular music (such as Talei Burns) and ultimately the supremacy of God in these islands (see Horst's discussion of Vodafone advertising). Vodafone has also astutely invested in charitable trusts supporting local community projects and, more recently, instigated the localisation of the company by accepting the 79 per cent stake of the National Provident Fund, the country's main pension scheme. All of this combines to secure the semiotic links of Vodafone to the place of Fiji and its future trajectory. In Horst's view, the moral economy that Vodafone promotes is focused on citizens more than consumers in the citizen-consumer conjugation.

Digicel, which launched later in October 2008, tried rather unsuccessfully to break this conjugation between the brands of Vodafone and the Fiji state, first by challenging the monopoly on the link to rugby but also by changing the palette of their brand to have the blue of the Fiji flag echoed in the last letters of their brand name. They also vaunted the fact of their entry into the market through liberalisation as a manifestation of freedom and choice and, as in other countries, pitched to the rural poor through both their extensive coverage and their credit systems. (Vodafone first focused on urban centres but increased its rural coverage in anticipation of and in continued competition with Digicel). But there were some strategic slips in Digicel's branding: using visual images that offended the conservative dress codes of older Fijians (e.g. tank tops, swimsuits; see Horst, Chapter 4) and an undue reliance on Caribbean music and popular culture, rather evoking generic 'brown islanders' than specifically Fijians (Horst, Chapter 4). Perhaps the greater resonance of the global rather than the national in their brand might partially explain their pre-eminence in the youth and business markets. If this brand loyalty continues, they may expand their market share in the future beyond their present 20 per cent holding.[1] Horst considers that Digicel promotes a moral economy more attuned to people's identity as consumers than citizens in the conjugation of consumer-citizens.

Other chapters in this volume deal more intimately with how mobile phones have transformed experiences of place in specific locales: from the towns of Port Moresby and Goroka (Foster, Chapter 6) and Port Vila (Kraemer, Chapter 5) to the remote villages of the Sepik (Lipset, Chapter 1) and the Highlands and Western Province (Jorgensen, Chapter 3; Wardlow, Chapter 2). The passion for connection beyond the boundaries of embodied belonging to place and kin is manifest.

---

1   In situating the rival brands of Vodafone and Digicel in Fiji we might compare the national versus global resonance of the contesting Christian brands of Methodism and Seventh-day Adventism discussed so persuasively by Hiro Miyazaki in the context of Suvavou and beyond in Fiji (Miyazaki 2004; see also Tomlinson 2009).

Lipset records the keen desire of Murik Lakes' people to seek connections, using surrogate ladders and platforms and beach hot spots to do so. The most reliable hot spot is in an appropriately liminal zone, a doorway in a household where a private handset is made available for public use, for whatever personal calls are needed by other villagers. Lipset interprets this as an extension of a moral economy dominated by the gift and the reciprocity of kin – most calls being about familial health and illness or about coordinating movements of kin, neighbours and business partners in time and space. The mobile phone palpably allows a compression of both space and time.

Wardlow (Chapter 2) discerns a rather different pattern in the Tari Basin of Highlands Papua New Guinea whereby mobile phones enable a novel connectedness between kin separated not just by geography but by gendered power and proprieties. Mobile phones enable women to keep contact with natal kin and, especially, mothers, sisters and daughters in a way that virilocal residence and husbands' proscriptions inhibited or even impeded in the past. But, beyond these reanimated kin connections there is also, for both women and men, the new phenomenon of intimate strangers known as phone friends (see below). Phone friends also loom large in the mobile telephone networks of young men in Port Vila, Vanuatu (Kraemer, Chapter 5). By Daniela Kraemer's account, based on fieldwork in the settlement of Freswota, young men feel defined by their urban settlement and increasingly detached from home islands and distant kin; the escalating costs of inter-island travel has made travel within the archipelago far more difficult than in the past.

Jorgensen's chapter extends the spatial reference beyond the national and the regional to the global. Using the case study of a man called Toby,[2] he shows both the paradox and, indeed, the deep ambivalence about the ubiquity and daily necessity of the mobile phone. Toby is suspicious of the immoral or even Satanic potential of mobile phones and the capacity of the 'mobile system' to yield universal surveillance and, thereby, the expanded efficacy of global superpowers like the United States and China. This may be a personal and even an idiosyncratic view, combining Biblical stories derived from the Book of Daniel with fears about undue foreign influence in the extractive industries of Papua New Guinea and the spectre of world government. But Toby is not alone in his fears. In both Papua

2   All names used by authors in this volume are pseudonyms.

New Guinea and Vanuatu, mobile phones are regularly associated with moral ills, many of which are thought to have foreign origins and are seen as signs of the 'end times' (Kraemer, Chapter 5; Taylor 2015).

Time

Several chapters allude to how mobile phones are intimately entangled with dominant narratives about, and the embodied experience of, time. Jorgensen points to how mobile phones have become linked to widely circulating apocalyptic narratives about the end times, 'a low background hum in PNG popular culture', which are especially pronounced in evangelical Christian and HIV contexts (see Wardlow 2006). In several chapters in this volume, and in previous studies, the labile potential of the mobile phone has been seen as a conduit or even a cause of loosened sexual morality, of illicit love, of rape and sexual harassment and of sorcery (see Andersen 2013; Taylor 2015). All of these are condensed in widespread imaginings of the transformations of modernity through the enhanced spatio-temporal mobility effected by the mobile phone.

Robert Foster's chapter witnesses another temporal pulse at work: the time-discipline inherent in pervasive prepaid schemes and transfer of credits that enable access to mobile phones by those who are resource poor. Inspired both by the earlier writings of Thomson on the moral economy of time thrift in factories and textile mills and the more recent writings of Schüll on Las Vegas casinos, he suggests that the time discipline encouraged by prepaid schemes assumes individual fiscal responsibility but simultaneously solicits excess consumption. The combination of using only what one has already paid for and the mandatory expiry dates of SIM cards, credits and alluring bundle deals leads to a logic of 'use it or lose it' and of stimulating subscribers 'to consume more in less time'. So, rather than mobile phones being primarily instruments of pleasurable, leisurely communication, their use becomes rather the disciplined work of consumption (evoked in the singular case of Lucy). As Foster suggests, 'The notion of time thrift … seems perverted when people stay awake until the wee hours of the morning to make phone calls'. The author of a letter to a PNG newspaper asked whether Digicel thought subscribers like him were nocturnal, like flying foxes. Foster also shows how Digicel's marketing regimes underscore the pervasive practice of seeking and giving phone credits: through using missed calls to say 'call me' and through the ease of sending credit to another user. Credit requests might be refused (suggesting avoidance or resistance), always agreed to or, more often, selectively agreed to in a way that signals and secures familial, friendship and romantic attachments. But, in all this, the company's regimes and personal dispositions interact so that, rather than being a respite from work, performing personal and social relations is a form of 'work' in Papua New Guinea in the era of late capitalism.

Person

Finally, I want to turn to questions about whether the pervasive use of mobile phones in these three countries has been influential in transforming notions and values of the person, from an emphasis on more relational to more individual forms. Lipset's chapter addresses this question most directly, pondering whether the voices of its subscribers are modern, rational and individuated. Mobile technology has the potential for greater individual autonomy, privacy, anonymity or even outright dissimulation. For instance, Andersen (2013) has written about *gesfaia* (Tok Pisin for phone calls made by anonymous, primarily male, callers to unknown numbers) as experienced by women students at a nursing school in the PNG Highlands. Such anonymous phone encounters, like the Internet, allows space for individualist fantasies about identity and 'lies and trickery' on both sides, even if those involved eventually become long-term phone friends. Lipset claims *gesfaia* 'suggest the rise of a decidedly Melanesian form of individualism, one that is free of social constraints imposed by kin and foreground the ambiguity of contemporary urban life'. However, Lipset also argues that the new connectivity is informed by the moral economy that predates it: 'mobile phones expand the physical, but more importantly, the moral, space of kinship'. For the people of Murik Lakes, mobile telephony 'expands space and time informed by normative, kinship-based values and the moral economy of the gift'. The space-time compression characteristic of modernity and, particularly, mobile telephony thus simultaneously generates a new cosmopolitan individualism *and* the maintenance of place-based collectivities between rural locales and urban diasporas. Lipset's analysis is exemplified by Venis and Eric's 'accidental kiosk' in the doorway of their house in a Murik Lakes village, whereby their handset was not treated as private property but freely shared with many others in the village, primarily to enable and perpetuate the density of kin relations at a distance (see above, and see also Lipset 2013, 2017; Telban and Vávrová 2014).

Kraemer's portrait of the use of mobile phones by unemployed and poorly educated young men in Port Vila suggests a rather different scenario (see also Kraemer 2015). The young men with whom she works are marginalised from both *kastom* and commodity economies, lacking both the resources generated by access to the land and kinship of home islands and the cash and commodities accessible through paid work or business. Even though they are barely eking out a living, they spend much of their time and scarce money on relentless quests for receiving and spending phone credits. For them, mobile phones are 'less about communication with family and friends than it is about broadening one's network of social relationships'. This extends to making random calls and sending texts to anonymous strangers in the hopes of accumulating more 'phone friends' and creditors. Such expansive networking can, as in the case of Alfred, become burdensome – too many contacts, too many texts – and thus the need to focus selectively on more valued contacts.

As Kraemer discerns, these dynamics are modelled on entrenched 'Melanesian' idioms of how generosity and strategic gifting yields status. But novel dimensions emerge in this situation of urban precarity: rather than re-inscribing existing relations of place and kin connection, these young men develop more egocentric, individualist networks of cashed-up kin and neighbours, politicians and young women who evince a romantic interest. Giving and receiving phone credits signals hope of developing relationships, of positive future acts of generosity, hospitality and loyalty on the part of the other. But a cessation of such reciprocity signals the devaluation or even termination of a relationship, often and most graphically in romantic attachments. The scenario of revenge and retaliation that developed between Toby (on behalf of his friend) and a young woman, Rose, reveals how phone credit exchanges and refusals are ways for young men to perform power and masculine potency in novel contexts, at a remove from models of relational persons embedded in place or kinship.

Wardlow's chapter evinces graphic evidence of how mobile phones are as much affective as effective technologies. Like Lipset (Chapter 1) and Andersen (2013), she attests to how mobile phones are used in heterosexual romantic liaisons between strangers at a distance, in which both partners can relish teasing, flirting and the potential for dissimulation, as well as sharing intimate stories of their daily lives. But her focus is on the phone friendships cultivated by women living with HIV, for the most part mature, middle-aged women. Such friendships typically originated in random anonymous calls during which the HIV-positive women did not hide their status but openly declared it to their phone friends, and shared with them intimate stories about their daily lives and, in particular, their sufferings not just from illness but from stigma and familial or communal exclusion. These phone friends thus became intimate strangers with whom it was possible to share troubles and trauma and to gain solace and support, but who were also at a safe distance. Without the 'thick relations' of kinship, often congested by anger and anguish, women could be easily affectionate – talking and laughing and relieving the stresses that they were told might reanimate the illness. The contraction of the social world experienced by many HIV-positive people was thus expanded again through such individual networks. The theft of a phone or the loss of the numbers of crucial phone friends could, as in the case of Lucy, be a source of anguish and deep personal loss.

In conclusion, this volume delivers innovative insights into the recent, short life of mobile phones in the Pacific and offers interesting comparisons with the broader, vastly accumulating literature on Asia, Africa and the Caribbean (see, for instance, Doron and Jeffrey 2013; Horst and Miller 2006). They have obviously had a profound impact on social relations in these three Pacific states, reconfiguring relations of place and time and the prevailing character and values of personhood. This is likely to continue into the future. This set of essays is distinctive in the way that, through a series of grounded ethnographies, it connects the political economies and regulatory regimes of mobile phone companies with how mobile phones are being used in everyday lives. The chapters by Horst and Foster reveal the strategies used by companies like Digicel and Vodafone to create an aesthetic of brand cultures, linked to ideas of the nation and modernity (through consumption as much as citizenship) and how they model systems consonant with prevailing patterns of use and subscriber desires. But the ways in which mobile phones permeate everyday lives and the moral worlds of subscribers far exceed these strategies, in conversations between distant kin and intimate strangers; in secretive, quotidian confessions and distantiated, dissimulating fantasies; and in imaginations of both utopic and dystopic futures presaging the 'end times'. The social life of mobile phones in the Pacific palpably reveals how they are not just an instrumentally effective technology but also a much-desired 'affective technology'.

This is part of a wider pattern of commodification of desire in late-capitalist culture:

> the emergence of brands and branding cultures moves beyond the shift to commodification to offer immaterial products such as emotion and affect as well personalities and values, factors that have become essential to the inner workings of global capitalism. Through these branding activities, users are transformed into consumers and citizens, or consumer-citizens, who use the products of a particular company and come to identify with the brand that produces or distributes a particular product. (Horst, Chapter 4)

But, as Horst suggests, this is not just based on the contracts between companies and consumers but also depends upon the relationships constituted with the state and the character of the regulatory environment:

> the cultivation of consumer-citizens is as much about the semiotics of emotion and imagery of belonging used in advertising as it is about ownership structures and regulatory bodies. Notions of national belonging are thus intertwined with telecommunications infrastructures. (Horst, Chapter 4)

The state personified is thus not just an inert background but an active agent in competing moral economies involving companies, consumers and the state.

References

Andersen, B. 2013. Tricks, Lies and Mobile Phones: 'Phone Friend' Stories in Papua New Guinea. *Culture, Theory and Critique*, 54(3): 318–34. doi.org/10.1080/14735784.2013.811886

Doron, A. & R. Jeffrey 2013. *The Great Indian Phone Book: How the Cheap Cell Phone Changes Business, Politics and Daily Life*. Cambridge, Massachusetts: Harvard University Press.

Horst, H.A. & D. Miller 2006. *The Cell Phone: An Anthropology of Communication*. New York: Berg.

Kraemer, D. 2015. 'Do You Have a Mobile?' Mobile Phone Practices and the Refashioning of Social Relationships in Port Vila Town. *The Australian Journal of Anthropology (TAJA)*, 28(1): 39–55. doi.org/10.1111/taja.12165

Lipset, D. 2013. *Mobail*: Moral Ambivalence and the Domestication of Mobile Telephones in Peri-Urban Papua New Guinea. *Culture, Theory and Critique* 54(3): 335–54. doi.org/10.1111/j.1548-1352.2009.01031.x

—— 2017. *Yabar: The Alienations of Murik Men in a Papua New Guinea Modernity*. Cham, Switzerland: Palgrave Macmillan.

Miyazaki, H. 2004. *The Method of Hope: Anthropology, Philosophy and Fijian Knowledge*. Redwood City CA: Stanford University Press.

Taylor, J.P. 2015. Drinking Money and Pulling Women: Mobile Phone Talk, Gender and Agency in Vanuatu. *Anthropological Forum*, 25: 1–16. doi.org/10.1080/00664677.2015.1071238

Telban, B. & D. Vávrová 2014. Ringing the Living and the Dead: Mobile Phones in a Sepik Society. *The Australian Journal of Anthropology (TAJA)*, 25(2): 223-38. doi.org/10.1111/taja.12090

Tomlinson, M. 2009. *In God's Image: The Metaculture of Fijian Christianity*, vol. 5, *The Anthropology of Christianity*. Berkeley: University of California Press.

Wardlow, H. 2006. *Wayward Women: Sexuality and Agency in a New Guinea Society*. Berkeley: University of California Press.